The Protection Formula (thinking like a cop)

by Lyle Arnold, B.A., C.P.P.[1]

Philippi Publishing
P.O. Box 23317
Pleasant Hill, CA 94523-0317

Edited by Paul Lippman, Pro-Edit
P.O. Box 160 • Sedona, AZ 86339

ISBN 0-9650241-7-2

Library of Congress Catalog Card Number 95-72860

Illustrations by Mary Arnold, Bernadette Arnold, Bernard Arnold and friends of the Arnolds.
Idea for cover design by Dixie Robertson of Griffin Printing, Sacramento, California.

Printed by:
Griffin Printing & Lithograph Co., Inc.
4141 North Freeway Blvd. • Sacramento, CA 95834

Cover design by John Chater
Front cover photo by:
John Chater Photography • Berkeley, California

Book design and layout:
Stornetta Publications
P.O. Box 2821 • Sacramento, CA 95812

To Marty, my wife, and our children Theresa, Veronica, Daniel, Nicholas, Vincent, Jennifer, Mary, Bernadette, Lyle III, Bernard and Charles; and their children, Rebecca et al. May they be protected.

PROLOGUE

"It's the great myth," said Kilvinsky to Gus, ignoring the voices behind them, "the myth whatever it happens to be that breaks civil authority. I wonder if a couple of centurions might have sat around like you and me one hot dry evening talking about the myth of Christianity that was defeating them. They would've been afraid, I bet, but the new myth was loaded with 'don'ts,' so one kind of authority was just being substituted for another. Civilization was never in jeopardy. But today the 'don'ts' are dying or being murdered in the name of freedom and we policemen can't save them. Once the people become accustomed to the death of a 'don't,' well then, the other 'don'ts' die much easier. Usually all the vice laws die first because people are generally vice-ridden anyway. Then the ordinary misdemeanors and some felonies become unenforceable until freedom prevails. Then later the freed people have to organize an army of their own to find order because they learn that freedom is horrifying and ugly and only small doses of it can be tolerated."

The New Centurions, Joseph Wambaugh
A Del Book, N.Y., 1970, p. 86
Reprinted with author's permission

TABLE OF CONTENTS

INTRODUCTION

A perennial debate exists as to whether politics is an art, or whether it is power[2]. This question is very pertinent to this book, because I have always believed that cops live lives curiously similar to those of politicians: that a cop's life is a microcosm of a politician's life. They both act out their jobs in an arena of conflict, where they must constantly discern the nature of their jobs. Especially important to this problem is what you will soon learn as 'the #1 conflict'. This is the type of conflict where physical action has not yet taken place. There is always a propensity, strong for some, less-so for others, to propel a #1 conflict into physical action. For the politician this means war, for the cop it means a battle. Both politicians and cops sometimes push over the brink what could have been resolved by talk. Both, that is, use power instead of art. What follows in this book is a view based on the belief that 'thinking like cops' for protection is an art.

It would be hard to find a better introductory story for my thesis than the Hollister, California, motorcycle 'riots' of 1947. The main reason for this is because a movie was made of that incident which the reader can review. The title of the film is "The Wild One" starring Marlon Brando.[3]

The real-life story of the incident took place on July 4th and 5th of 1947, when about 3,000 'bikers' converged on Hollister, then a town of 4,000 (with seven police officers), to participate in the Independence Day festivities, which included motorcycle speed contests and hill climbs. Among the motorcyclists was a group known as the 'Booze Fighters'[4], which triggered the day-and-a-half rumpus.

An *ad hoc* force of 29 cops quelled the trouble by noon of July 5th. There were some arrests for such violations as indecent exposure, some traffic fines and a few county jail sentences for as much as ninety days. But when Hollywood got hold of the story, it fictionalized, romanticized, and otherwise distorted the true event, with this distortion no doubt contributing to the fringe attractiveness of 'motorcycle outlaws' everywhere.

In the Hollywood version, which is mainly of interest here, the Black Rebel Motorcycle Club, led by Brando, roared into town like a cast of young hawks checking out a peaceful, virgin countryside. After gassing their cycles, they began frolicking, first in the streets, then in a local restaurant and bar.

There was conflict among the townspeople from the beginning. The sale of beer, food and gas incited business owners; the radical novelty of the club's members attracted others.

Some, however, clashed with the crude swashbucklers. Then a rival motorcycle club arrived, fights broke out, and revelry and drunkenness began. From that point on, wrongdoing escalated. Laws were broken, and there was only one condescending and awkward cop to try and maintain control. Vigilantism took hold, and more laws were broken, both by the vigilantes and the outlaws. Finally, a man was killed.

The value of this movie lies in the ultra-graphic sequence of *escalation*, from good natured frolicking, to minor law violations, to moderate violations, to serious felony violations resulting in manslaughter (with a vigilante committing the latter). Book One is about precicesly this escalation — and the sequel need for protection — as *should* be seen through the eyes of a cop.[5]

Cops protect themselves better than most people because they *think* differently. They've been programmed to think and respond in a certain way. Their ensemble of thought patterns are a cut apart from those in commerce, industry, academia and other professions. Like others, whose way of earning a living causes them to face danger regularly, they must think in such a way that will get the job done and stay alive. Combat soldiers, firemen, prison officers and others likewise rely on a unique mind-set for efficient work production.

But thinking like *cops* is especially important to today's citizen in this century as we approach the year 2,000, because every day cops must do two things to be proficient in their duties. The first is stay alive, and the second is avoid legal problems. With today's violence, all citizens must do this. All citizens should develop the habit of thinking and *acting* like cops because we all live in a dangerous society. Until this happens, there will be no control of crime and violence in America.

The scope of this book is purposefully narrow. Only one motive and one objective were in my mind when I sat down to write it. The motive was generated by my daughter being kidnapped, and almost murdered[6]. The objective followed: to teach her and others the art of protection, by offering the reader a short course in how to *think* in a dangerous situation, in order that his or her defense would come *ab intra* (from within) rather than *ab extra* (from without).

When a person acts from within it comes off with more potency. For example, have you ever listened to a person *read* a speech? That's a speech from without. We've all been inspired, however, by a speech from within, the standing ovation kind, like the famous trial lawyer Clarence Darrow used to deliver. I was moved in reading them, even though I *disagreed* with Darrow's way of thinking!

To accomplish this objective, I drew upon a *Formula* I had created and used for thirty years, first as a prison officer, then as a law enforcement officer, and finally as a security executive for a Fortune 500 corporation[7]. Book I is about this Formula (and, tacitly, so are Book II and Book III). Each element in the protection chain is covered, step-by-step, from the least dangerous situation, to the most dangerous; from verbal arguments to bloody violence, from heckling to butchery, the conflicts all cops cover in their careers.

In the chapters which follow, I shave off the excess of security instruction, and 'write its rules and principles on the head of a pin'. To accomplish this, I made this work a *little* book. I attempt to jump-start the reader into *thinking* about protection in our dangerous society, and then to read more about the subject from the myriad of available books (some of which I list in the following chapters.)

> If you read the newspapers and watch the news on TV...
> If you are a thinking person and see the danger in America...
> If you care about your well being...
> If you care about the well being of your loved ones...
> This book is for you.

The United States is being deluged with crime and violence, while at the same time "the proud American citizen cowers at home, worried sick about crime and delegating, through votes and opinion polls, ever more power to 'anti-crime' measures that increase statutory police power and rely upon lock-down procedures."[8]

Crime and violence is the primary concern of citizens in California — more than the economy, more than public schools, more than AIDS[9], and it is the number one problem in the minds of Americans generally[10]. The chances are higher than ever before in our history that you and your loved ones will become the victims of this crime and violence. In fact, one report shows that "5 out of 6 Americans will be victims of violent crime"[11]

It may have already happened to you or your family. If so, this book will prepare you for the next time.

If it hasn't happened yet, consider yourself lucky, and begin preparations now, by following the short steps outlined in this book, to sharply reduce the chances of becoming a victim, though you can't eliminate them altogether, even if you move to the woods.

With the present crisis of crime and violence in America, this Formula is now presented to the citizens of America for the first time, that he or she may think like a professional in times of danger. This book could save your home and family, your fortune, your job, your well being, your life. And it could do the same for all those who depend on you.

Recent news flashes:[12]

- Driver shoots motorist to death because of high beams
- Softball pitcher kills man for heckling him
- Man kills his neighbor because neighbor's dog urinated on his newspaper
- Teenager kills boy for staring at him
- Woman kills person for complaining about her cigarette smoke
- Tourists being killed in 'guerrilla jungles' of America
- Criminal justice system in U.S. only a minor inconvenience for criminals
- Comatose, quadriplegic woman in hospital raped, now pregnant
- Driver murdered for driving too slow in fast lane
- Deaf woman shot in face for using sign language (gang member mistakenly thought it was a sign from a rival gang)
- 38-week-old unborn infant 'ripped' from mother's womb with scissors and a knife

The reader undoubtedly knows this list could go on for page after page. The point is that we live in dangerous times, and all thinking people are experiencing a warning. A violent storm is brewing, and it could break at any time, on the streets, in your home, in your business.

Some people experience the dangers in modern America from the news, others from living in mean neighborhoods, and others because they or their loved ones have already become victims.

How about you? Do you feel it 'can't happen to me?' If that's your attitude, this book isn't for you.

If, however, you make it your business to be prepared, read on. The book you are about to read is based on the simple idea that: 'Wars are never won by pessimists'.

A study of the early 1980's found that 75% of Americans fear to walk in their own neighborhoods at night[13]. Pessimism. To get through a dangerous situation you must be an optimist[14], and to be an optimist, you must be ready for whatever comes at you. You must be prepared. Just like the old Scout motto says, 'Be Prepared'.

Now, with the dangers facing the citizens of America, it's overdue that everyone should possess the simple but effective strategy of the Formula. When it becomes part of your everyday thinking, you will be different than you now are — you will be an optimist.

Whether you're a commuter, a housewife, or a student, the contents of this book will change the way you view yourself and your loved ones in a dangerous society. It will give you a positive, rather than a negative outlook for thinking ahead.

Planning for protection:

Nobody goes to work without a plan, even if that plan is embedded into the memory through routine. A housewife instinctively knows what her movements will be in the morning, in the afternoon, in the evening. A carpenter visualizes the structure he will build, be it a frame for a wall or an entire house.

A dentist listens to the aches and pains of the patient, probes, takes X-rays and goes to work. A student plans each and every day of the chapters to be read, the references to clear up the unknown, the notes to be taken, in order to learn.

All these people (everyone, in fact, who has a task to perform), works according to some kind of formula. When it comes to facing *danger*, however, most people get caught up in a web of tension, uncertainty and confusion, for the simple reason it happens so quickly; they are taken by surprise, and it's a new ball game for them — they're not prepared.

Dread can be the outcome, and panic. This dread and panic can and *does* happen to unseasoned cops. But they develop, over time, internal mechanisms to deal with the dangers.

Then, in a few years, maybe five years, maybe seven years, they are seldom surprised by the danger. They have had a series of successful encounters, and they become 'programmed' to recognize and meet new circumstances. Weed recognition will demonstrate the point.

Tribulus Terrestris:

In California, and perhaps elsewhere, there grows a horrible weed. Its botanical name is *Tribulus Terrestris*, and its common name is the 'puncture vine'. It grows prostrate on the ground, with stems proceeding outwards from a hub, forming a 'wheel' with a diameter of about sixteen inches. On the vines are small oblong leaves, and ghastly kernels of needles and spears somewhat resembling a medieval mace. These little maces are the bane of bicyclists and joggers. They will puncture a tire or tennis shoe with the utmost severity and impunity. Once I had seven puncture holes in my front tire from running over one! Flat tires are common on bike trails, but a patch kit doesn't have enough patches to handle much of this kind of weed attack. The bicyclist must, out of necessity, become watchful for this weed. In the measure he doesn't he will contine becoming a victim. He will reach a point, however, when he becomes programmed to recognize this weed in time to avoid its wrath, in much the same way a cop recognizes *beforehand* what he is up against.

Such a 'program' you will now learn about recognizing danger ahead of time, for your protection. The steps are all so simple that the reader might well say, 'I could have thought of this Formula myself!' And that may well be true.

Once this is done, it can be incorporated into everyday living, like a simple formula for making a cake or learning how to type. The difference here, however, is that you won't get a chance to *use* the formula every day. People not in the profession of danger will keep the formula *latent but active*, which can be done by simple imagination: one minute a day, two minutes a day, whatever time it takes you to keep the Formula alive and well.

You'll be assisted in this endeavor merely by reading or hearing or watching the day's news of the latest atrocities of crime and violence. And you will imagine how *you* could have protected yourself or others in each situation.

By doing this, the stories of crime and violence will lead you into a learning process that is positive and fun. Learning is actually fun if it is directed to an endeavor which will benefit you. And the benefit you will now learn can be life saving.

THE FORMULA

My Formula is based on an analysis that, whenever protection happens (except by chance, luck or interaction), three things, or rather three catagories of *events* have come into play; and three types of conflicts are kept within their parameters.

I. The Three Events

Category 1: The person had a certain state of mind.
Category 2: The person understood certain things.
Category 3: The person had arranged certain things.

state of mind/understanding/arrangements

S / U / A

Catagory 1: *The state of mind*:
We begin with the notion of *fear* (rightly understood). We can't face danger properly unless we are afraid. This fear is not a servile fear, but a *healthy* fear — there is a profound difference between the two.

> "...soldiers die in largest numbers when they run, because it is when they turn their backs to the enemy that they are least able to defend themselves. It is their rational acceptance of the dangers of running that makes civilized soldiers so formidable...Men fight... from fear: fear of the consequences first of not fighting (i.e. punishment), then of not fighting well (i.e. slaughter)."[15]

And when this healthy fear is operating, it moves the instincts to a tension which can be likened to a batter at home plate. His body and stance are poised for the strike, his eyes, brain and nervous system are integrated, as the ball is flung towards the plate. He is ready.

It is precisely this fear, and the tension it generates, which allows us to observe exactly what it is that threatens us, and to meet the danger realistically. That is why our rightly-understood fear must be a constant in our lives. It must be habitual, like locking up the seat belt. If we fasten the strap only half the time, the crash will surely come in the other half, when it was hanging loose. The same with danger. It happens from one moment to the next, very often with little or no warning, so our healthy fear and benign tension must be a part of us, like eating and sleeping.

Tension doesn't happen by itself. Something causes it to come into being. And when it arrives, it causes the second principle to come into play, that of becoming *alerted*.

And, once alerted, we become *prepared*. Preparation doesn't happen in a vacuum; it happens because something is telling us to take action. The mother who admonishes her child to be careful crossing the street is *afraid*, as she well should be. So here we have category one, the mental state: there was f*ear*, the person was *alerted*, the person was *prepared*.

Obviously, if Mary had no fear for Jack's safety, she would merely send him off to school each day with the false optimism of fools.

Catagory 2: *The understanding*:

When Mary instructs her son Jack to cross only at crosswalks, to watch the traffic, to be alert, she also has a certain *understanding*. She knows that the danger to be focused on is the trip from home to school.

An actual time and space format is embodied in this trip of potential danger — five minutes to walk the two blocks to school. The danger period is enclosed in that time of five minutes, like a phrase within a sentence is enclosed in a *parenthesis.* There's a beginning and an end to this parenthesis.

Mary also places this danger within a *perimeter*, so Jack understands in terms of space where the danger is. The perimeter here is the space between home and school. That is the immediate concern of the mother. There's other dangers she must prepare Jack for, with both larger and smaller perimeters, but she takes them one at a time and makes sure he understands each, always using the *parenthesis* and *perimeter* (time and space) concepts, to make things clear. With these concepts Mary is able to isolate each security problem. She is able thus to delineate her separated security concerns, and keep the minor concerns subordinate to the major concerns. In other words, she is able to prioritize them. And Mary knows that the object of her fear is the enemy, who we shall call the *viper.*

The *viper* is the threat against mankind. It is fire, water, earth and man

ENEMY SCHEMATIC

The viper isn't always an enemy, but when it turns dangerous, its threat must be *estimated.* There has been a kidnapping of a child from Jack's school recently, so Mary's estimate that the viper could strike her child is a prudent one.

Category 3: T*he arrangements*:

A. *Communication.* Mary communicates with her son: 'Beware of strangers'. 'Cross at crosswalks'. 'Watch the traffic'. She would like to buy him a cellular phone, so he could always communicate dangers, but she can't afford it, so she tells him that some strangers are very bad. Mary further arranges for

B. *Transportation.* Jack only has two blocks to school, so the arrangements here are for walking. Longer distance, different mode, but transportation is always a factor in our Formula.

C. *Tools.* Mary has given Jack a small, but very loud whistle. A child running from danger and blowing a whistle (or screaming) will attract anyone's attention.

Recap of the three events:

1. FAP: Fear, alerted, prepared. The state of mind.
2. PPVE: Parenthesis and perimeter, viper, estimates. The understanding.
3. CTT: Communication, transportation, tools. The arrangements.

Every cop, on every beat in the country (once he or she becomes a veteran), has these three events embedded in the bones. He fears the guy who can hurt him with a healthy fear, which keeps him alerted and prepared at all times. The time and space of his beat becomes second nature to him; he isn't worrying what's happening in the next town, he's concentrating on the time and space of his beat. He knows the viper is there, which is his reason for existence, the reason he gets paid. And through the years he has learned to make estimates of the enemy's strengths and weaknesses.

And when the viper comes, he's ready with a communication system for getting help, and he has transportation arranged at all times, and he has tools (a gun, handcuffs, a notebook and more). And it is because a cop has these three events working, that he knows protection better than most people. He thinks like a cop.

II. The Three Conflicts

The conflicts you will now learn complement the three categories of events. They can be listed under three headings. Cops face all three of these conflicts, and develop an instinct for each. The first type involves noise and trivia, the second involves moderate threats to person and property, and the third involves serious threats to person and property.

1st Conflict: noise and trivia (soft conflicts)
2nd Conflict: threats to person or property, moderate (tempered conflicts)
3rd Conflict: threats to person or property, serious (hard conflicts)

These three conflicts can be depicted as follows:

<table>
<tr><td rowspan="2">Threats from noise and trivia

1</td><td colspan="2">Threats against person and property:</td></tr>
<tr><td>tempered
2</td><td>hard
3</td></tr>
<tr><td>Softly resolved by tactics</td><td colspan="2">Resolved by physics: i.e. matter and energy (force)</td></tr>
</table>

The vital faculty a cop has, that the lay person doesn't, is *quick* recognition of what kind of conflict he or she faces. This perspicacity is what allows the cop to deal professionally with most any given situation.

The lay person, on the other hand, usually has no immediate perspective of *exactly* what he or she is dealing with. There probably is some perspective, but the lines are blurred. And this obscure picture, or this incomplete picture, can get the lay person hurt or killed. To be ignorant of the dividing lines of these three catagories, or to ignore them, has caused ruin for countless citizens, and the termination or worse for legions of cops. I cannot be accused of pedantry on this point.

Rodney King, Ruby Ridge, Waco. These incidents caused elaborate investigations, criminal trials or hearings on whether or not the dividing lines were crossed by those in authority. The use of force is clearly spelled out for law enforcement officers on these dividing lines in precepts on the effects of force, reasonable force, deadly force and illegal force. Yet in the three incidents mentioned above, all handled by trained, professional officers, many experts both in and out of the law enforcement community believe the dividing lines were violated. This confusion, about what kind of force is acceptable in what kind of situation, has clearly reached high levels. It doesn't matter what someone thinks, or feels about this problem. What matters is what public opinion and the law will *allow*. We don't live in an age when people who are out of line always get put in their place. I started in law enforcement in 1961 — three years before Escobedo, and five years before Miranda[16] — and we did things with impunity then, that can't be done now. Cops in 1951 did things I couldn't do, and cops in 1941 did things they couldn't do. It may not be right, but that's beside the point.

You *must* stay in the conflict category you're in! If it's a #1, you can't use the force that's required in a #2 or a #3. However, if it's a #2 or #3 you can't treat it like a #1. To coldly recognize these three conflicts, to acknowledge these lines, and pay them honor is vital to health and welfare. It once saved my job in a serious police brutality suit.[17]

This Formula (the three events and the three conflicts) is what this book is all about. It can be expressed as:

S. U. A. + 1 | 2 | 3 = **PROTECTION**

If you learn it, if you master it, you will *think* like cops do, and *act* like them, for protection. They have the Formula somewhere in their constitution giving them the wherewithal to do the right thing. They have the mindset, the tools, communication and transportation, and the conviction that they must treat with precision the three types of conflicts, in order to avoid legal hot water, and save their lives.

In a book about warfare, John Keegan sought to "reduce the conduct of war to a set of rules and a system of procedures, thereby to make orderly and rational what is essentially chaotic and instinctive."[18] My intention is the same. The three catagories and three conflicts can indeed become 'chaotic', and therefore the Formula seeks to make them 'orderly and rational'.

The method of learning has already been stated. Every time you see or read about violence in the news, or even see it in a movie, imagine *yourself* in that situation applying the Formula. It only takes a minute. By constant repetition of this, your life and the lives of those dependent on you, will become more secure from the nurturing that is about to begin. Good reading and maximum protection.

BOOK I

PERSONAL AND HOME PROTECTION

PART ONE: HINDERING THE NUISANCE

Chapter 1: Handling noise and trivia

> *"God made the angels to show him splendor — as he made animals for innocence and plants for their simplicity. But Man he made to serve him wittily, in the tangle of his mind!" Act Two, "A Man For All Seasons"*[19]

Sometimes we can't stop the process of verbal antagonism, so we must use our *wits*, that is, we must act 'wittily in the tangle of' our minds. For guidelines, we shall turn to The Formula. All of Part One (two chapters) are spent on a discussion of pre-*physical* confrontation tactics. We call this the art of soft conflict resolution.

It is the most common type of conflict and it happens to everyone. It's soft, because if handled softly you can stop the escalation into a more serious conflict, a harder-to-handle problem. However, even though it's soft, it may be the most difficult of all conflicts to handle. Soft things are sometimes deceiving, like quicksand can be.

The conception of a problem:

John and Ted are neighbors — but in name only. They are both of the middle class variety, both shower every day, both take care of their responsibilities, both have red blood coursing through their veins without distinctive amounts of alcohol or other chemicals therein (most of the time anyway) — but they hate each other.

The conflict started several years ago. Maybe it was John's kids playing in the backyard, and throwing things into Ted's yard that started it. Or maybe it was Ted's dog barking at John's kids. The issue has become clouded over the years as to the exact incident which began to hemorrhage. But now days it hemorrhages regularly with shouts, threats, insults and occasional blood-shot eyes and bloody noses, both resulting from high-pitched yelling and escalating blood pressure.

For the umpteenth time the beat cop arrives, and knows exactly what he will hear. He knows the exact timbre of John's voice inflections, the exact pitch and intensity of Ted's assertions, and the exact story each will tell him.

"Who can live with that dog barking? Hear it? It sounds like some kind of paranoid Tasmanian Devil! Could you live with it, officer? Listen to it!" John demands.

"And why is White Fang barking? Huh? I'll tell you why!" answers Ted. "Because one of his brats threw a handful of Screaming Yellow Zonkers over the fence! That's why! Tell me you could live with that officer! Come on, let's hear it! That dog's just doing it's job!"

The scene develops predictably, with nobody happy (far from it), least of all the cop.

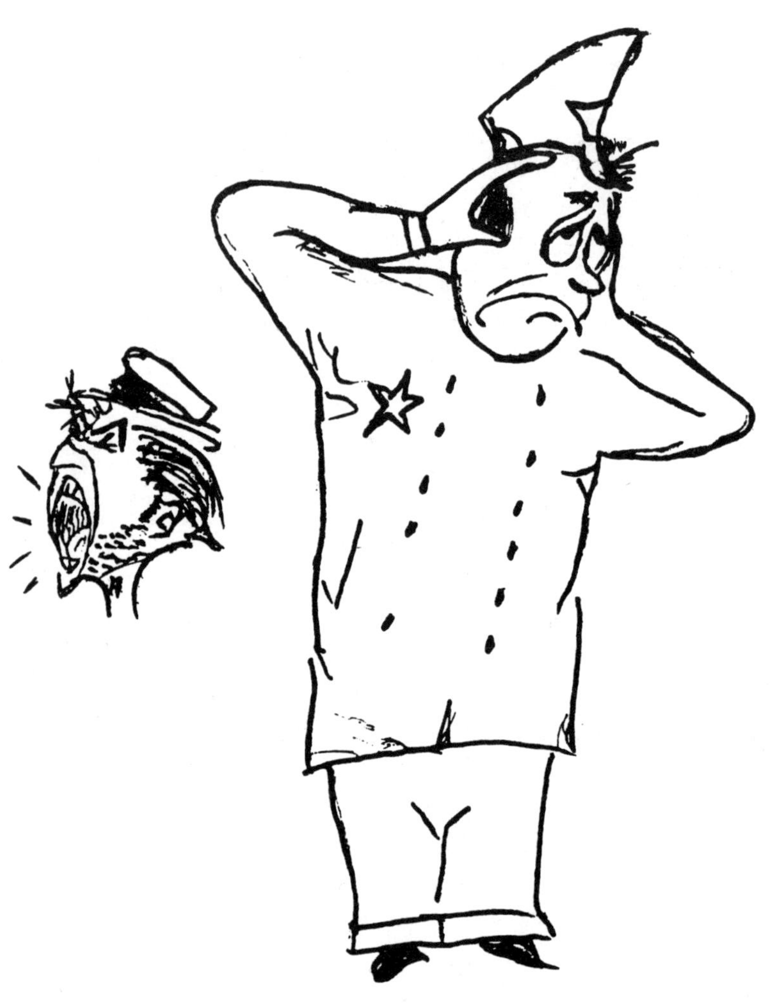

He would prefer making a good arrest, or having one of the local pukes flip him off. He simply didn't get into police work for this kind of crap. And both John and Ted now hate each other more than ever.

John has thought seriously about a clandestine operation, where he sneaks into Ted's yard when nobody's home, drugs (or even poisons) the dog, puts it in a plastic bag, and throws it into a dumpster. Ted has contemplated having a 'hospital job' done on John, something he saw in the movies. A guy in a bar, in fact, mentioned once to him that he knew a 'bulldozer' who would guarantee two broken bones — a hundred dollars for each bone. Cheap at twice the price. Both think about their vendettas regularly; both are truly tempted. Both really aren't kidding.

Both are on the line between a soft conflict, and a conflict which spells real trouble: taking hostile action against another's property or person, maybe less than death-threatening, but, the nature of escalation being what it is, maybe not. They don't think of themselves as bad guys, just as standing up for their rights, 'like anyone would.'

Both are closer to danger than they realize, because they haven't learned the art of soft conflict resolution. This art begins with seven simple concepts:

1. Beast
2. HLOOA
3. Apex
4. Zone
5. Odds
6. Pace
7. RB/TJ

Concepts defined:

1. The beast concept. Like it or not, we all have two sides. Anyone who denies this hasn't been around. I knew a cop once who was always in control. Like perfect. No matter what the conflicts he had to handle (soft, tempered, hard), he always remained unruffled. Then one day after a bad shift, on the way home from work, an old lady tapped his rear bumper at a stop light. It was just 'a dime's worth of damage', but the cop got out of his car, walked to the lady's car, climbed up on her hood, and jumped up and down on it for about ten seconds. It almost cost him his job. It was his first time up, however, and he was allowed to get counseling and take some time off (most of it without pay), and it worked. He was ready for his beast the next time.

The beast, however, isn't the real problem. The real problem is the confusion the beast causes. Confusion, one of my professors used to say, is not when you don't know the solution; it's when you won't recognize the problem. And the problem here is to admit you have two sides, one of them capable (given the right circumstances) of acting like a beast. And mastering this animal is possibly the hardest job in the world. Literally, because it means mastering one's self.

Few of us accomplish it fully. Because we trust ourselves, we look out for ourselves. Our interests, our knowledge, our attitudes, all these redound to that which is most important in the world: to our well-being, to our respect, to our honor.

And this is good. It is natural and right. In fact, it can't be otherwise. So *behold* the reason such a dilemma exists, for handling the conflicts of noise and trivia! Two natural elements come together in a clash: an annoyance vis-a-vis our self interest. All that remains (after we admit we have a bad side, which must be brought under control), is to learn the next step.

2. HLOOA. Head lights out of adjustment. When a driver approaches you with his high beams on, it is not only aggravating but dangerous. When a traffic officer stops a driver for this violation, the first thing he usually does is to look at the dashboard to see if the high-beam indicator is working, or defective. If it's working, he will often cite the driver, because the driver should have known. However, sometimes it is not the high beams at all, but headlights that need adjusting. In the latter case, a driver can drive for months without realizing he is blinding other drivers.

   Now many people are like the driver with his headlights out of adjustment. Many people aggravate, and even insult other people, without realizing their faults. They *should* know what they're doing, but that's not the issue. They really and truly don't. The driver who drives with his headlights out of adjustment for months on end, has had a hundred drivers blink him to tell him he's blinding them. But he simply doesn't get the message, and keeps on tooling along without fixing the problem. The guy, or gal, who goes through life aggravating (and even insulting) people, like the guy who drives with his headlights out of adjustment, has to be treated with courtesy when correcting and admonishing. It's the best way. If *anything* will work, that's what will work, being courteous. But this admonition needs qualification.

   In institutional life, such as the military, academia, and prison systems there are sanctions against insubordination. It would be impossible to keep order without them. A teacher cannot let a classroom brat dishonor her without penalties. The same is true in other institutions. Action must be taken, but it should be taken with courtesy. The victim of insults must not enter into the arena of the offender, and lower his or her position or dignity. It 'pays off in spades', as the saying goes. But this book is about thinking like *cops.* A street cop does not have the controls of institutional life. When he is insulted he can't send the perpetrator to the principal's office, or write up a citation for a higher authority to deal with. He is forced to retain his authority, with all its attachments, both abstract and corporeal, on the spot. How does he do this, and remain courteous? He does it like Michael did in Mario Puzo's THE GODFATHER, when he was insulted by Moe Greene, the former 'Murder Incorporated' thug, with Michael's brother Freddie remaining obsequious.

   > "Freddie, you're my older brother, I have respect for you. But don't ever take sides with anybody against the Family again. I won't even mention it to the Don."

He turned to Moe Greene. "Don't insult people who are trying to help you. You'd do better to use your energy to find out why the casino is losing money. The Corleone Family has big dough invested here and we're not getting our money's worth, but I still didn't come here and abuse you. I offer a helping hand. Well, if you prefer to spit on that helping hand, that's your business. I can't say any more."

There are a thousand variations to this tactic, but the first principle always remains the same: detachment. You can't control either yourself or others if you're attached to your antagonist through emotional perturbations. *Always* remain detached from your subject.

3. The apex concept. At the Highway Patrol academy we learned the art of high-speed driving. Fully helmeted, and strapped in like an aviator, with roll bars on the cars, we were placed on what essentially was a race course. There was a 440cc engine under the hood. My instructor was a former race car driver.

   I had already spent five years in police work when I switched jobs to the Highway Patrol. In the former department I had been in several high speed chases, yet we had never been taught the *principles* of high speed driving. I marveled that I was still alive, after being taught these principles.

   In many endeavors — maybe even most — there is a simple *something* which must be grasped, in order to master this endeavor: one thing, the *soul* of that endeavor. For a bricklayer it might be that certain twist of the wrist, as the mortar is swiped across the bricks. For a secretary, it might be the ability to type only with the fingers, with no thought patterns. In high-speed driving it is the apex of the curve (the inner-most point).

   It's very simple. You merely head into a curve at below maximum speed, and, as much as possible, hug the apex of the curve. You must *accelerate* as you round the curve, which causes centrifugal skidding. Kinetic energy carries the car outward, but this 'powerslide' (as it is called), is controlled, and if proportioned, the car will not leave the roadway. Whereas if the driver enters the curve from the middle of the roadway, or worse, towards the right side of it, he won't have the space for the controlled skid. Further, if he goes into the curve too fast he must brake, and — unlike accelerated skidding — this kind of skidding can't be controlled. With the uncontrolled skid, the danger is very real that the tires will break traction and move over the line, off the roadway.

So important is this concept for handling noise and trivia, that it would be impossible to overstate its importance. Take for instance a conflict of verbal confrontation. If we go into this dispute 'like gangbusters', we get carried away by our own inertia. Then we become part of the problem, instead of the solution.

By going into the verbal exchange below maximum, we can keep our emotions intact. Just as the high-speed driver knows that he will jump the boundary of the curve if he accelerates too fast, so the master of handling noise and trivia knows that he will plunge into dangerous waters if he lets his emotions run too high. He lets his opponent skid out of control. He goes on the defensive, in case his opponent makes this dive into troubled waters. But he stays afloat on his raft of good intentions. He goes into the dispute quietly, guardedly, accelerating gently, and hugs the apex of his emotional curve. He doesn't lose control.

What often throws the noise and trivia conflict into tilt is the belief that the exchange of ideals is a good in itself. This simply is not true. Sorry to be a heretic, but nothing could be further from the truth. If there are controls on this exchange, it is a good in itself. Like in a courtroom, where a bailiff with a gun doesn't let things get out of hand. In this case facts are separated from opinion, inconsistencies in testimony are exposed, the truth is found, and someone pays the price for being wrong.

But if there are no controls — beware! A novel with the title UNCLE TOM'S CABIN propelled our own country towards its deadliest war. Words can kill. Alexander Hamilton, one of our founding fathers, discovered that words can be deadly. At a dinner party in 1804, he called the vice president, Aaron Burr, 'dangerous'. Burr killed him for it. At that time the duel was the form of honor which settled such things. But don't kid yourself. The duel is still alive and well. People kill each other over parking spaces in this country.

Go into the curve below maximum. Accelerate in controlled amounts as you enter the curve, and hug the apex of the curve so you don't skid out of control. Keep the beast in its place; don't let it push 'the pedal to the metal'. Be at war with yourself, win the victory against self, so you can resolve the conflicts in life.

4. The zone concept. Alice's son Robby, though good at heart, sometimes wanders over the line in his young teenage quests. He once was caught shoplifting, but Alice was able to settle with the store without police action. Another time he got drunk on 'sneaky peek' wine (and rumor had it there was marijuana involved), and threw up on the coffee table, ruining an expensive decorative book and a new set of leather gloves. His language, also, sometimes is atrocious. She has to constantly bawl him out for using obscenities, and for yelling at his siblings.

   In both the stealing incident and the throwing up, Robby was with his new friend Norris, and Alice is convinced it is Norris' influence that is leading Robby astray. He was expelled from school twice, and has been in juvenile hall three times. And Robby's language and temper were more in order before the two of them started palling around together. Her endeavor is to split the two up, and get Robby with his old friends, all of those decent kids. She simply has no *control* over Robby when he is with Norris, no control at all.

   Control is a big word in conflict resolution. When there is conflict, some components of this conflict can be placed under our control, and some components *can't* be. To recognize the difference between the two is vital: what we can control, and what we *can't* control.

   In Alice's case she was lucky. She had no direct controls over Norris, but she could take steps to eliminate his influence over her son. Many times in life, we are not as lucky as Alice.

   During many noise and trivia incidents, we simply can't control the other side of the problem. It is not within the *zone* of our control. If a street bum spits out disgusting remarks at you for not giving him the coins he asked for, this is beyond your zone of control. It's happened already, you were simply at the wrong place at the wrong time. You could return the ridicule perhaps, or slap him around, but there's imminent risk in both.

When I was a prison officer, I read with great interest once in a convict's file an account of his gang's operation. Several of the gang would conceal themselves, and one member would accost a man walking with his wife or girlfriend. He would make deliberately lewd and insulting remarks towards the female; and when her mate moved on the criminal from pure rage, the entire gang would close in and rob, rape — and sometimes kill — their marks.

The street bum may not have a gang waiting. But he has very little to lose in any encounter. In fact, he may be waiting for an incident with police involvement, to spend a day or two in jail, get a shower, some hot meals, and rest up for awhile. But you'll be taking your own time to go to court, and if you did slap him around, that's assault and battery. You'll lose.

5. The odds concept. Not many years ago the odds would have been maybe ten percent that, in a noise and trivia scene, your opponent would opt to escalate the conflict, no matter what you did. Today, in the mid-nineties, I would give it much higher odds, maybe 30%. Even higher, depending on where you live. We live in troubled times. There are people on the streets today who won't give in, no matter what the cost to them in time, money or even health. Many of the protagonists you meet "don't care about *themselves (*italics are mine), and they sure don't care what they do to you."[20] Studies, in fact, as far back as twenty years ago have shown that up to forty million of our citizens are psychopaths.[21] If that was true in the early 1970's, what is the figure in the mid-1990's? Today our social circuits are overloaded, and a very large number of American citizens are zapped out.

   Accept it. Don't fall victim to a noise and trivia incident. Use the ideas you're getting in this book to avoid it. You've got better things to occupy your time.

6. The pace concept. When you can't avoid a noise and trivia entanglement, *pace* yourself, until you can free yourself from it altogether. There are nine innings to a ball game. Fifteen rounds in a title fight. Four quarters on the gridiron. And in fencing, when one side scores five hits (or pokes, if you will), he or she wins.

   If you must play the game of noise and trivia, play it like any other game, and know that you probably won't get a fast win, and maybe no win at all. In fact, it's better not to try for a win at all in noise and trivia incidents, at least in the common meaning of the term. In noise and trivia games, you win if there's no escalation.

   Think of yourself as a contestant in one of the above sports, trying to play the sport only for a given duration. *Pace* yourself, it may be a long game. Or think of the various gaits of a horse. He can walk, do the fox trot, or the regular trot, or the pace (a two-beat gait, like a waddling), or the rack, canter, or gallop. Try and keep the gait as slow as possible. Trot if you have to, but don't gallop.

7. The RB/TJ concept. In the movie classic "Gone With the Wind", Rhet Butler is insulted and about to be challenged to a fight, simultaneously, by the impulsive patriot, Charles Hamilton. RB walks away from his antagonist without losing any dignity whatsoever. He didn't have to put Hamilton in his place. Hamilton's fit of words was seen by the bystanders for what it was: impulsive and irresponsible.

   How *many* of the bystanders understood that Hamilton, and not Butler, was the loser? This is never known. But this 'Charles Hamilton syndrome', as I think of it, is a fact of life all of us must face. People are free in this country to be verbally abusive, at least to a point. Therefore, all of us at times are on the receiving end of this injustice. Our lot is that of Rhet Butler's; we are stuck with it, if we ourselves are to be responsible. If we are not to be the agent of escalation in verbal confrontations, we must do like Rhet. But the dilemma is not as awkward as it may seem.

   Already covered is the notion of fear (rightly understood). This is exactly what RB had. If he would have succumbed to the whims of Hamilton, and gotten drawn into a fight, he would have had to kill him. And Butler easily could have done this, but he chose the lesser of two evils. He walked away. He didn't want the trouble all this would have caused him. And we can do the same. Here's a variation of the RB concept, taken from the novel, GRAPES OF WRATH, by John Steinbeck. It is an argument that was quickly escalating to a physical fight, until Tom Joad (TJ) sought to divert it:

   > "If I pay you half a dollar I ain't a vagrant, huh?"
   >
   > "That's right."
   >
   > 'Tom's eyes glowed angrily. "Deputy sheriff ain't your brother 'n law by any chance?"
   >
   > "The owner leaned forward. "No, he ain't. An' the time ain't come yet when us local folks got to take no talk from you goddamn bums, neither."
   >
   > "It don't trouble you none to take our four bits. An' when'd we get to be bums? We ain't asked ya for nothin'. All of us bums, huh? Well, we ain't askin' no nickels from you for the chance to lay down an' rest..."
   >
   > 'Tom was silent for a long time. His dark eyes looked slowly up at the proprietor. "I don't wanta make no trouble", he said. "It's a hard thing to be named a bum. I ain't afraid", he said softly. "I'll go for you an' your deputy with my mitts — here and now, or jump Jesus. But there ain't no good in it."
   >
   > 'The men stirred, changed positions, and their glittering eyes moved slowly upward to the mouth of the proprietor, and their eyes watched for his lips to

> move. He was reassured. He felt that he had won, but not decisively enough to charge in. "Ain't you got half a buck?" he asked.'
>
> Tom Joad had just been released from prison for squashing a guy's head with a shovel, during a fight at a country dance-hall. He wasn't the kind to walk away from a fight. But here he had too much to lose. He had his family with him, so he gave his opponent a loophole. This loophole wouldn't have existed if Tom had gotten carried away. Tom knew they were in a #1 type situation, and didn't exceed the boundaries. We all can take a lesson from Tom.

For generations Americans have been exposed to the 'cowboy mentality', where the good guy takes no flak and always wins. Don't make Hollywood decisions on the streets, or anywhere else. They can be very costly. They can cause loss of health, financial difficulties, even financial ruin. And in some cases they can result in injury and even death.

Chapter 2: Avoiding Escalation With A Good Feeling

Physicians of old used to believe there were four types of people running around, each with a different temperament (or 'humor' as they were then called). Each type responded in a different way to a given stimulus — a person's remarks, something they read, an unruly child.

The belief at that time was based on four fluids in the body: blood, phlegm, yellow bile and black bile. The person controlled by blood was a *sanguine,* the phlegm person was a *phlegmatic,* the black bile a *melancholic* and the yellow bile was a *choleric.* Though intelligence ranged in each group in the normal way, here's the differences in today's parlance:

1. Sanguine person: cheerful, easily excited but doesn't stay mad long, not deep thinker about problems — takes them as they come.
2. Phlegmatic person: doesn't get excited, makes calm decisions.
3. Choleric person: quickly (and deeply) responds to stimulus.
4. Melancholic person: doesn't respond readily to stimulus, but if it's repeated enough, this type becomes similar to the choleric.[22]

Look out below!

Emily's the pillar of the community. Being a choleric type, she dedicated herself at a young age to being a perfect example of wife and mother. Her role model was an aunt, whom everyone called a 'saint'. People come to Emily with their problems, and she always seems to have time for everyone's burdens, family and friends alike. And even though she gets infuriated when confronted by unfairness and bigotry, most of the time she controls her temper, for a greater good, and most of the time her spirits are high.

Today, however, she got some unsettling news during her regular medical checkup. They want to do a biopsy, and the doctor was guarded (or so it seemed to Emily, because she thinks so deeply about everything). And, to make matters worse, she has to have a root canal tomorrow. Worse yet, one of her teenage daughters has been dating a man considerably older than her (and there's a rumor he's married), all of these problems causing her gnawing pain and worry. On this day, it seemed, everything was in erratic motion. It was the worst day possible for Kitty to be acting up.

Kitty lived on the same block as Emily. Their relationship was always cordial (mostly because of Emily's toleration, painful for her at times), but never friendly. Even the name 'Kitty' bothered Emily. It sounded cheap. And Kitty was of the sanguine variety. Sanguine through and through. Not a serious bone in her body, and no intellectual giant she.

It happened so quickly. Emily was on the phone with her daughter, who wanted to stay out all night 'with a girlfriend'. Emily was horrified, almost to the point of nausea. Her fear —

actually it was closer to consternation — seemed to swell and burrow into the bad tooth. It was painful, to the point of bitterness. It thumped in her jaw, just as the fear of the 'married man' and the biopsy were thumping in her brain and nervous system, reaching every cell of Emily's body and soul.

BANG! Sounded like a gunshot! Emily told Katherine, the daughter, to hold while she ran out to see what the noise was. Her heart was in her throat, things were not in the right perspective anyway, and this could be the crowning blow. She was sure one of her own kids had been shot. They lived in the inner city, and drive-by shootings were getting almost common.

There stood Kitty on Emily's doorstep, laughing. "Thought I'd get a rise out of you on that one" joked Kitty, honestly cheerful, delighted, in fact. Her son was a track coach, and she had his starter gun (which used blanks) in her hand. Dumb move to be sure, but, for Emily, the wires holding her brain and body together snapped all at once.

She came through the screen door like a football lineman practicing on a field dummy. She was in white heat. She grabbed the starter gun and whacked Kitty across the face with it, splitting the cheek with a deep laceration. Then she threw the gun at a car parked in front of the house at the curb, and drug Kitty by her wrist to the sidewalk, flinging her into the same car. She was screaming words at Kitty she never used before.

The cops were there in no time. The criminal and civil action which followed took well over two years, and almost cost Emily her marriage, family and home. This was especially sad because her biopsy was benign, and her daughter really *was* spending the night with a girlfriend, because she'd spilt up with the 'married man' (who actually wasn't married at all). And the root canal went smoothly, and healed up in no time. But something good *did* come of the situation.

Emily passed from being the neighborhood hero to being what she'd always aimed at: a 'saint'. The two and a half years of 'exile' that was spent during the criminal and civil action caused Emily to humble herself more and more, until she could relate to the sufferings of others more than ever. And now it didn't matter at all to her if the person who needed help was a simple person like Kitty, or a genius. Having gone through a trial where she was wrong, she now understood that 'everyone makes mistakes'.

Emily, in short, felt good because she understood how life really is. She saw — in a way she never thought was possible — that she had another side, with a beast in there, a beast that can take control even in *noise and trivia* conflicts. And though costly, she used this newly found knowledge to brighten her life, rather than drag it down.

She had gone over the line from a noise and trivia incident, to a much more serious affair. But now that she knew the rules, she could help others even more. She became, in a very real sense, wise. She realized that anger is only good if one gave way to it in conformity with reason, propriety and honor. But this is, by definition, restrained anger, not the kind she got carried away with.

She taped up on her cupboard something Aristotle said, that *we hate all those who oppose our impulses.* She often mused after her ordeal that overcoming whims and caprices is a lifetime job. She understood clearly that the endurance which must be sustained in physical trials often is much easier, simply *because* they are physical. It's the control on the *inside* that is hard, because it involves the will, in its most pristine and assertive form. We all want to be right; we all want to win. But we can't, and so we have to learn to endure a type of interior pain when we're around others. More for some, less for others. The conflicts of mankind begin on the inside, be it with the advice of a Machiavelli, or the interior rage of a Hitler, or the vicious paranoia of a Stalin.

Challenge from a fool:

There is a species of antagonist in the #1 we can lable the 'PITS'. He or she is always *persecuting* someone (P), always *intimidating* someone (I), always *taunting* someone (T), always uses *sarcasm* (S). The perpetrator of this type of noise and trivia species seeks to exert a superiority over you, a grip on you, and to maintain this grip. Sometimes the antagonist has a messianic complex, sometimes it is to punish you for something he thinks he sees in you. There are many examples.

Ramona was an avid bicyclist, and lived in an area where bicycling was common. On one stretch of roadway where there used to be a bike path, a construction area caused the bike riders to use the sidewalk for about a half mile. It was a narrow sidewalk. One day she was pumping up hill on this sidewalk, and a rider was zooming down the same sidewalk. He was taking up almost the whole sidewalk. As he passed Ramona he screeched out at her to "Use the other sidewalk!" They almost *collided*! It seemed he was trying to force Ramona off the sidewalk, and there were cars whizzing by in the lane right next to her!

It took a couple of minutes for it to register with Ramona, just exactly what this other rider was up to. He had an idea. The sidewalks on this construction stretch should operate like traffic lanes, one direction for each lane. The sidewalk on the *other side* of the street should be used for riders going up hill, 'his' sidewalk is for those peddling down hill. He was creating law!

The only realistic resolution for PITS incidents is to keep written records of them. Then, if they persist, use your documentation to get peremptory action. This could be from a superior, through legal channels, or even in the arena of public opinion. The important thing is to handle it systematically, and not let it turn into an ongoing feud.

PITS people often are in violation of the law, especially misdemeanors like 'disturbing the peace'. In California (the law for other states is often the same in substance, but check your's out), the Penal Code makes it a crime for:

(a) Any person who unlawfully fights in a public place or challenges another person in a public place to fight. (b) Any person who maliciously and willfully disturbs another person by loud and unreasonable noise. (c) Any person who uses offensive words in a public place which are inherently likely to provoke an immediate violent reaction. (section 415). Always try to avoid falling into a PITS incident.

The Tubby method of saving face:

When I was a youngster they had *Little Lulu* comic books. The male protagonist in the stories was Tubby. And once he got put on the spot by the town bully, a consumate PITS personality who tried to push Tubby into a battle.

Tubby told him, "Okay, but I get to pick the weapons." The bully was a scrapper. He wasn't thinking of *weapons.* But he couldn't say no, because then *he* would have looked yellow.

Tubby then told him to meet him the next day on a certain vacant lot, and he would provide the weapons there. The weapons were two buckets of tar, with a broom stuck in each tar-bucket. They would cover each other with gooey *tar*!

The bully backed out, as he knew his mother would wail the daylights out of him if he came home covered with tar. Tubby had the right instinct. Get the other guy on *your* turf. Don't be dumb enough to fall into a PITS trap. The other guy may be able to out-class you, two to one, in his field of expertise.

Someone once said you should never challenge a fool. We can modify this saying to read as follows: 'When challenged, if you accept, you may become a pitiful *fool*'. The bully thought Tubby was a fool. He knew he could have cleaned Tubby's clock, hands down, in a regular fight. But Tubby used his brain. He got the bully on *his* turf.

We're supposed to be *thinking* creatures. The world is full of pitfalls, traps, decoys, nets of all kinds. We're more than muscle and blood and bones — we have brains. Let's use them. When a PITS person accosts us, it's a trap. Noise and trivia are perennial ambushes. Using your intelligence and your wits isn't being yellow, it's being human. And the more we adhere to our nature, the more we feel good.

Three ways to walk away from trouble:

1. When you're right, and the other side's wrong. Don't fulminate. Don't make a person feel like a fool, even when he or she has acted the part. Always let the other guy keep his dignity. You've been wrong before, and will be again.

2. When *you've* goofed, and the other side treats you like a disgusting animal with no right to being. Think about what's happened. He flipped out because you screwed up. Now are you going to make it a match? There are a thousand ten-second counts to bring yourself under control, to avoid the coming explosion. Once I screamed out in Japanese the ten count: itchi, nee, son, chee (our D.I. used to count cadence in Japanese). Once I took out a handkerchief and chewed on it until my frothing and drooling was absorbed by the

cloth, and my sanity was returned. Once I jogged at full speed on city streets for an hour, in a business suit, after having a rabid argument with a superior.

3. When your opponent transmogrifies (turns into a monster), during an otherwise normal exchange: find a way out. Some people can transmogrify in a pico-second, right while you're talking to them. Here's a personal experience that could have gotten me killed: I was on a narcotic investigation, working with an ex-con (who was my confidential informant). We were in a bar in an inner city neighborhood, when unexpectantly a person who was once in prison with my informant recognized him. He sat down, and I was introduced as an ex-con who also had known the informant in prison. As we chatted, it became clear (I thought) that the newcomer was carrying a firearm. It was even more clear that he was quite likely psychotic. I had an instant brainwave, to see if I could get the serial number off his gun, and get him picked up (he was a parolee). I was carrying an attractive foreign automatic, and went into the men's room and removed the clip and the chambered round. Then, at an opportune time, I slipped it into the ex-con's hand under the table. I thought he'd compare his piece to mine, and I'd get the number off it, or at least it's description. But he was no longer the same person. He became something out of a Stephen King thriller. My informant and I had to wring the gun out of his hands. Something in this guy's system caused all his planets to line up at once when he touched my pretty gun. He lusted after it, needed it. I was sure, as he got more crazy, that the incident might gum up sixty days of hard work, so I eliminated my body from the picture altogether. Luckily my informant was smart enough to help me pull it off. I walked away from something I wanted to confront.

Training for victory over the noise and trivia incident:

To be prepared for any confrontation, we must be *prepared.* One doesn't get into the ring without a good set of lungs from roadwork, and without lots of swinging and bobbing and weaving in basic maneuvers. So, too, with other confrontations: we must train. Here is a little mental gymnastic you can do to train for the noise and trivia clash: Picture yourself at a four way intersection. It is controlled by signals on all four sides. You are in a hurry. You're late for something that's important to you. You are stopped at a red light, and there are four cars in front of you, and several in back of you. The seconds click by...a minute passes, then two, then three. Long light. If things go right, and if you hustle, you still have time to make the appointment. After the fourth minute at the signal it registers with you that it's stuck. Horns are blowing. There's nothing you can do, because you're boxed in. The car in front of you takes a chance and makes a U-turn, but a car coming around the corner collides with it. Pandemonium. However, you now have a clear spot to pull out, and as you begin to accelerate your car dies. You turn the engine over, but all it will do is whine. The guy in back of you honks his horn at you, and yells a familiar epithet out the window. You're about to blow it.

This happens to all of us from time to time. And there's no better way to depict a noise and trivia scenario, when you're already well on your way into the trap. Simply speaking, you shouldn't be where you are. It's almost too late to stop the claws of the trap from squeezing you into a mistake. But, with practice, you can backtrack to the time when you're rushing to get to your important appointment. All of us get into this predicament, sometimes often. Often it's poor planning on our part. But even if it's not that, we must resign ourselves to the fact we live in an imperfect society, and we're not perfect, so these things happen. We can't have perfect control over mishaps. But we can control the first step when it happens.

And, by controlling the first step, we can control the whole series of steps when we get caught up in noise and trivia explosions. When we get pushed, when we get intense, we can have a talk with our better self, and conclude that if we miss the appointment, it isn't the end of the world — far from it. But if we let ourselves get fired up with the notion, 'Nothing is going to keep me from this appointment. *Nothing*!', then we are sitting ducks for escalation. This practice, allegorically, is a perfect one for mental training for the noises and trivias in our lives. If we train ourselves, we will softly neutralize these nuisances, and have better lives.

PART TWO: SHIFTING DOWN FOR PROTECTION

Part One was the soft — and often the most difficult — part of the protection Formula. Now we must shift down, and gear ourselves into a slower speed. We'll encounter more dangerous situations here, and we don't want to move too fast if we're going to handle them properly. This second kind of conflict is of the *tempered* variety.

When we temper something, we bring it into a state where we can use it to our own benefit. This is done by mixing something with the raw product, or treating it in some way. Thus we temper paints with oil, or temper steel by heating and sudden cooling, or temper criticism with reason.

The tempered conflict can be easier to handle than the soft kind, because sometimes it's easier to 'let it go' than hold it in. The big difference is that tempered conflicts, like the even more dangerous ones (hard conflicts), must be dealt with by some kind of energy, often mixed with matter. 'Force' is another word for it. This, of course, is not to say that all cases of tempered conflict will go unresolved without the use of force. It merely means that we must be *prepared* to use force.

Professional conflict resolvers, such as police, firemen and the military carry guns, or firehoses or other tools to neutralize the enemy. But just because a person isn't a professional —i.e. isn't getting *paid* to resolve a given conflict — doesn't mean the rules change. And the rule here is: to resolve tempered conflicts, we must be prepared to use force, either before, during, or after the fact.

Chapter 1: Using Force to Stop Minor #2 Conflicts

A. *Before the fact*

Beverly has a basic security system in her home. An intrusion alarm system, reinforced door frames, solid doors, good locks with deadbolts, smoke alarms, outside lighting, perimeter fencing. When leaving the house, she always sets the alarm, locks the doors and turns on the outside lights (which are activated by movement). She has just made the effort to resolve conflicts *before* the fact, by engaging energy and matter. Some of the energy, *she* provides, by turning the key, and flipping the switches. The alarm and lighting devices are examples of matter (and carry their own energy).

B. *During the fact*

Gwen's at the stove making dinner. A grease fire breaks out. On the wall in the kitchen is a Type B fire extinguisher. She grabs it and puts out the fire easily and quickly. (65% of all house fires start in the kitchen). Gwen has just stopped a conflict *during* the fact.

C. ***After the fact***

Ernie comes home from work. He'd forgotten to lock his front door. When he arrives, he notices his stereo set missing. (In about half the home burglaries in America, the thief enters *without* having to force his way in). Ernie picks up the phone and calls the police. He gives them the serial number on the set, and its description. Then he waits for the police force to recover his stolen goods. He has just acted to resolve a conflict *after* the fact. The only energy he applied was picking up the telephone. The police supply the force.

Chapter 2: Getting Familiar With Security

Security defined:

Security means protection. It means protecting something which is under attack, therefore it is a *defensive* mechanism. It must counteract an offensive where the attacker has chosen the spot for the attack, the method, the tools and other supports best suited, and the timing of the attack. The attacker also has made estimates of the strengths and weaknesses of his victim. Quite an advantage, all this. It can be seen why security must not be a passive state: it must counterattack. The prospective victim has to be *ready*. His security program must be *active*. The attacked must become the *attacker*!

In utilizing the Formula, it should be understood that it works at all levels of security, from personal protection all the way up to national security. This book deals with all levels.

The hierarchy of security:

1. The protection of individual life and property, or anything of value, including services and ideas.
2. The protection of state and local interests, institutions, business interests and schools.
3. The protection of the nation from foreign threats, including military. (Security is, therefore, a principle of war).

Chapter 3: Using the Formula For Home Protection

Melvin and Jennie Clark were stunned by the newspaper article: 'Child Dies in Swimming Pool'. The victim was the child of their neighbors, the Winston family. The Clarks also have a swimming pool and a child, and the newspaper article had pointed out that "Swimming pool drownings are the leading cause of accidental death for California children ages 1 to 4..."[23]. The *leading* cause! So they decided to develop a security system around their swimming pool, to avoid the same tragedy as their neighbors. They had never heard of the Formula, but because of the *fear* this death had caused in them, they thought long and hard about all facets of their problem, that they should be *alerted* if their child was in danger, and *prepared* for that danger. They knew this must be their state of mind, to be afraid for their child, alerted when she was in possible danger, and prepared for it — if it happened.

After reading of the tragedy, it took no genius mind to make *estimates* about how ruthless the enemy could be. Water can kill. They knew that the *time* needed to put in their security system would be well worthwhile, especially compared to the time it takes to drown. They knew they had to put a *perimeter* around the potential *viper*, so it would be contained. Their estimate of the danger would not be in vain.

They made arrangements with their work schedules to take care of the problem immediately. They *communicated* with an expert on swimming pool security, and he agreed to have his company *transport* in the necessary *tools* and equipment, and get started right away. A high fence was placed around the pool, a fence that couldn't be climbed by a child. A gate was installed with special equipment, a self-locking catch, and a very loud buzzer which sounded (*communicated*) when the gate was opened. This buzzer also sounded within the house, using speakers. (An 'On-Off' switch was installed, when the system wasn't necessary).

As the reader undoubtedly has seen, all of the three (SUA) steps for the Formula have been enacted:

1. FAP: Fear, alerted, prepared. The state of mind.
2. PPVE: Parenthesis and perimeter, viper, estimates. The understanding.
3. CTT: Communication, transportation, tools. The arrangements.

Achieving the castle mentality:

Gail, a single mother of two, was terrified. Right up the street from her a two-year-old child was kidnapped, from her *bedroom,* at night[24], while the rest of the family slept! She would become (unless found) one of 58,215 missing juveniles in the United States.[25]

"That's it!" she exclaimed. "I'm going to fix it so it doesn't happen to me!"

Gail didn't have the money to consult an expert on home security, but she did have a good imagination. She always had loved castles, and had a large picture of one in her front room. She began to study that castle, and to envision the medieval wisdom of security consciousness. "They may have had enemies at their walls," she mused, "but so do we!"

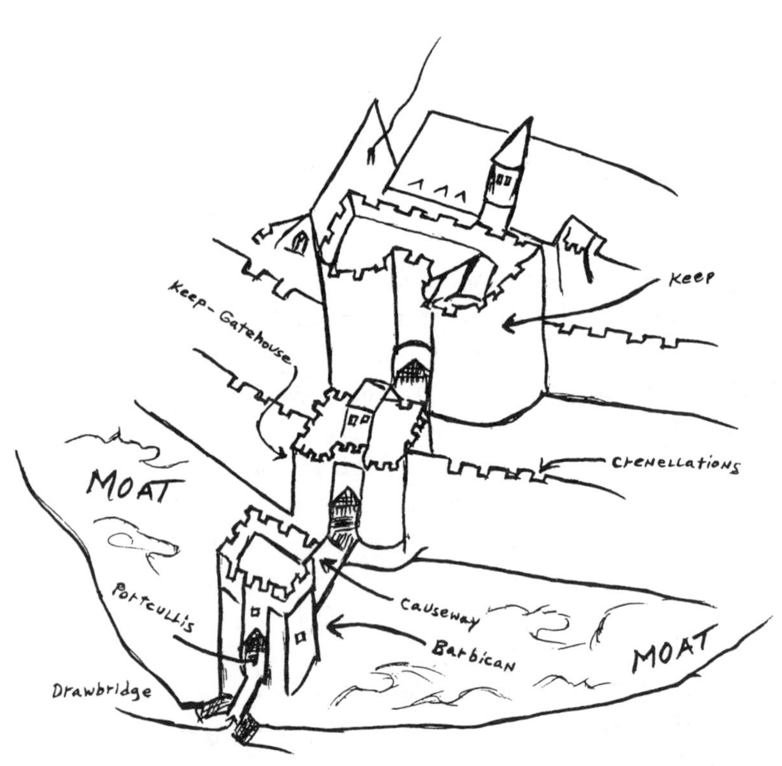

The first thing that impressed Gail was the use of various perimeter controls. A moat caused the enemy to devise a way to cross over water. High walls caused the enemy to climb. Together these controls caused the vipers a lot of trouble, and they would be vulnerable during their advance.

She also noticed the controls that were instituted in the openings to the castles. A *drawbridge* could be raised up, so undesirables couldn't even get to the front gate. Then there was a large *gatehouse* with protected walls (*barbicans*), and a *portcullis* (a heavy grating protecting the gatehouse, which could be lowered). There was a *causeway* (road raised above the moat), which further exposed the enemy as they crossed. And there were *keeps* (towers) for men looking out for enemies. Also the walls had *crenellations* (notches) for the defenders' protection.

Gail knew she couldn't have watchmen looking out for enemies, but she did have outside lighting installed, with a motion sensor, which would go on when there was movement. Surely this would give night-time intruders some trouble. She had grating placed on her windows[26]. She had the fence repaired around her back yard. She began a research for the right breed of barking dogs, and (being a lover of birds) even considered buying a Trumpeter[27]. She also would consider buying a 'Radar Watchdog'.[28]

She had a peep-hole installed for seeing who was outside her door. She had all her outside doors deadbolted, and solid doors and frames installed (almost 90% of breaking and entering occurs through a back or side door). And she had a loud, screeching audible-alarm installed that would sound outside her home during danger. To protect it from the weather, it was placed under the eaves. A 'panic button' (to activate the alarm) was installed in her bedroom, at entry doors and in the kitchen, and she had it put on line to an alarm monitoring station. In addition, she had smoke and carbon monoxide detectors installed at strategic points in the home, something she had been planning to do for months.

Finally, there were two vital items. Inspired by the *barbican* and the *portcullis* in the castle, she had a wall and heavy gate installed a few feet in front of her house. The wall was six feet high, running about thirty feet, from the side fence to the spot immediately in front of her front door, about twenty feet from the doorway. An iron gate was attached to the end of the wall, and locked onto a magnetic latch on the side of the garage. It was controlled by a switch inside her front door, and she installed an inexpensive communication system for talking to people at the gate. This gave her two perimeter controls around her house, the fence in the back yard, and the wall and gate in the front.

And lastly she devised a kind of 'keep'.

The keep is the innermost part of the castle, and the strongest. Gail's keep would be her bedroom. In case there was an intruder in the house, she had read that the resident should *never* go hunt for him. Instead she should barricade herself in a secure room and make a 911

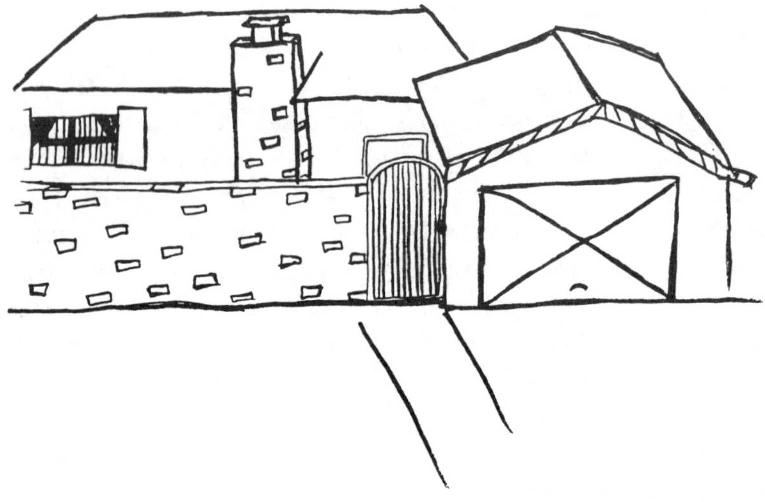

call. She also would hit her 'panic button' to try and scare him off. And she had her bedroom door replaced with a heavy door, tough locks, and a strong doorframe installed.

All this cost her less than a used car. She minded not at all the expense, when weighed against the tragedy down the street. And she made a commitment to herself to look into two other security devices.

She had read an article which told of day-time burglaries being very high. In this article it gave the results of a survey of prison inmates, whose business was burglary, which revealed that (except for being *witnessed*, or facing someone with a gun) the best protection was a burglar alarm with a direct link to a monitoring service, or, where allowed, the police department itself. She decided to talk with a local police officer about this, and also to contact an agency that installed and serviced residential alarms. Determined not to become the victim of unscrupulous salesmen, she decided to invest a couple of hours of her own research by:

A. looking into the standards for alarm contractor's installation (for households), set by the Underwriter Laboratories in Illinois. She found the address to be: Underwriter Laboratories, Inc., 333 Pfingster Road, Northbrook, Illinois 60062.

B. making phonecalls before signing a contract on any home alarm system. She would ask the prospective company to provide a list of their residential alarm users in her vicinity, together with their telephone numbers. By calling some of them, she could determine the quality of the service.

C. asking her local police or sheriff crime prevention units (or her insurance company) for handout material on home security, especially to learn about all the little things she may have missed; and to make inquiries about local alarm companies.

D. writing for reprints of Reader's Digest articles:
1. New Ways to Protect Yourself From Crime, June 1994,
2. Easy Ways to Protect Your Home, September 1994,
3. Riding With the Cellular Posse, October 1994.

E. writing to Butterworth-Heinemann, 313 Washington Street, Newton, MA 02158, 1-800-366-2665, and asking for their list of home security books.

F. reading Stephanie Mann's book titled *Safe Homes, Safe Neighborhoods - Stopping Crime Where You Live* (Nolo Press, Berkeley, CA).

G. writing to The Safety Zone, Hanover, PA 17333-0019 and getting a free copy of their catalog on security and safety products for the home.

H. *communicating* with her children about the tactics of would-be abductors. She will institute a pass-word. She will tell them that if she ever were to have a person pick them up, whether known or unknown to them, this person will know this pass-word. She will tell them about some of the *lures* abductors might use in approaching prospective victims, for example:
1. sympathy. "I hurt my arm, could you help me?"
2. bribery. "Would you like something really beautiful?"
3. directions. "Could you tell me how to get to River Street?"
4. authority. "I'm a policeman. You'll have to come with me."
5. injury. "Your mother has been injured. I'll take you to see her."
6. help. "Could you help me find my dog?"

And she would admonish them that sometimes the criminals merely grab the child, at which time they should blow their whistles very loud and do *anything* to get away, running ('break-and-run') to the nearest store, house, anyplace where there is an adult. And teach them how to fake a seizure, or just fall down and scream.

The second security device she was interested in came from a certain fear she had. Gail knew that things could go wrong with *any* system, *any* planning. She had been very impressed by the discovery of what went wrong during the 1986 Challenger shuttle disaster, which killed seven crew members. If her memory served, it was the putty (used to seal the 'O-rings' on the rocket booster) which caused this disaster. Billion dollar equipment and planning, she often thought, became a national tragedy because of faulty *putty.* And she had read about:

A. how a handyman had gotten into one of the most 'closely guarded places in' the United States: the War Room[29],

B. how a drunk youth had penetrated the security designed to protect the Royal Family[30],

C. how a burglar had broken into the private apartment of Prince Charles at the St. James Palace (which had a state-of-the-art security system), and 'lounged on Charles' bed, reading intimate letters'[31].

D. how a suspect named Andre Dallaire climbed the fence of Canadian Prime Minister Jean Chretien's estate at 2:23 a.m., on November 5, 1995. Three Royal Canadian Mounties were on duty and didn't respond to the alarm. The suspect wandered on the estate for 20 minutes, waved at the security camera, and then threw a rock through the Prime Minister's window. Carrying a knife, he encountered Chretien's wife, who barracaded herself and called for help. It took the Mounties seven minutes to respond.[32] (Author's note: 7 minutes is like an eternity in critical encounters).

So, finally, what if the worst happened? What if an expert housebreaker got in anyway? And what if he was a psycho? She remembered a recent news article, 'Why I Have a Gun'. The reasoning of the person who wrote the article was hard to dismiss. The woman who wrote it told of a man in her area, who got into houses and raped women at knifepoint. The writer said she got the gun for him. She said she practiced with it so she wouldn't shoot through a wall and hurt someone, but would hit the man instead. She said she wouldn't give the gun up until one of two things happened. "...either genetic engineers figure out how to grow women as tall and strong as men, or we all learn how to raise boys so that rape becomes unthinkable."[33]

Gail ponderd buying a gun, and decided to talk to someone she could trust about it, someone who cared for her well-being, and someone who would be objective, who was not prejudiced against guns. She felt much better now. She was imbibing the wisdom that came from facing enemies. Like her friends in the castle.

PART THREE: SHIFTING DOWN INTO 'GRANDMA'

When I was a teenager I drove a truck, and we used to call the lowest of the compound gears 'the grandma'. You could almost drive up the side of a wall with it. But it was very, *very* slow. That's how we have to think of what comes up now. When serious danger threatens us, the groundwork must have been laid out slowly, deliberately. In Part One we looked at soft conflict resolution, and learned that sometimes we can neutralize the conflict before it becomes tempered or hard. Then in Part Two we got a quick look at the use of simple force, before, during and after the fact; and got an overview of security. Now, with the foundation that's already laid, we can look realistically at a higher level of danger, including both the #2 and the #3 conflicts.

SECTION A: HANDLING NON-LIFE-THREATENING CONFLICTS

Chapter 1: How Not To Handle A Street Scene

Edmond was on his way to work, lunch-pail in hand, smile on his face, thinking of the night's bowling game. He had a sanguine character, and a moderate intelligence, but he was gifted with massive, powerful arms and hands. Ten years as a farm boy surely helped develop these.

As he passed by a group of teenagers waiting for the schoolbus, one of them commented about 'the gorilla'. Encouraged by laughter, the young man's insolence and boldness seemed to peak, as he blocked Edmond's path, giving him a moderate nudge with his shoulder (he was a lineman on the high school football team), with the comment, "Out of the jungle today, young monkey?"

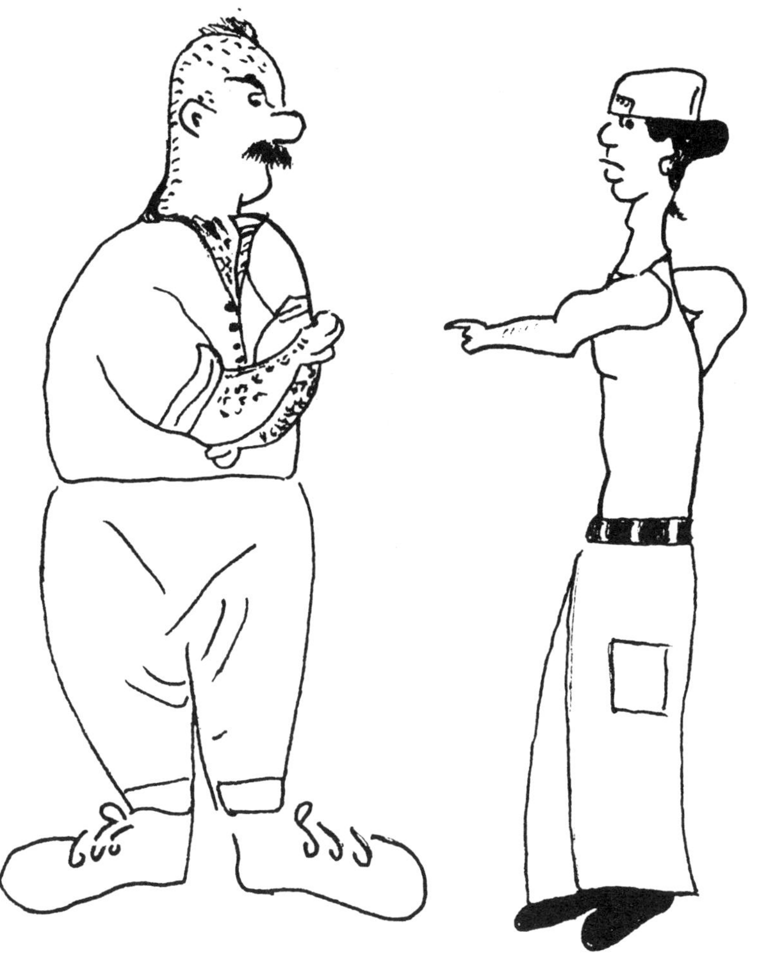

Ordinarily Edmond would have let the comment pass, as he was used to it by now. But a combination of the shoulder nudge, the snickering of the others, and the particular malice of the insult, made him more ticked-off than usual. Edmond turned to the youth, placed his two ruinous hands on the boy's forearm, and broke it. It was a clean break and a compound fracture, with one of the bones sticking out and covered with blood and muscle. Panic and nausea overcame the youth at once, and he went into shock.

Several of the youth's companions, infuriated, tackled Edmond and were badly beaten, one of them requiring plastic surgery, and all of them needing hospitalization. By the time the ambulances and the police arrived Edmond was sorry for what he had done. Nevertheless he was arrested, tried and convicted.

Considering all the circumstances, the judge went as easy on Edmond as possible. However, there were several successful and costly civil actions that followed. And with the criminal conviction and fine, and the stay in the county jail, it took Edmond years to pay for everything. In the meantime, he lost his job, went through a divorce, and became an alcoholic and a very bitter man. For the rest of his life he was in and out of jail for bar brawls, finally landing in the state prison for a manslaughter conviction, where his opponent during a brawl suffered a broken neck.

If Edmond had known the Formula, this all could have easily been avoided. He would have known that we live in a land 'planted thick with laws, from coast to coast'. He would have known that in our convoluted legal system (witness the 1995 O.J. Simpson trial, if a good example is needed), one easily could get caught up in a brier-patch of obstructive tactics, and the shifting moods which follow. He would have known that one has to be very, *very* careful these days when using physical force.

Chapter 2: Understanding The Scene

"Wherever law ends tyranny begins." 2nd Treatise, John Locke

CONFLICTS

1 /	2 /	3
soft	tempered	hard

In virtually all tempered and hard conflicts (#2 and #3 on our scale), there is a violation of the law. These are conflicts where someone has taken hostile action against another's person or property. In other words, you're dealing with someone operating outside the law, the tyrant, as Locke called him.

Now this tyrant may be only moderately dangerous (the #2 conflict), or gravely dangerous (the #3 conflict, covered later). The reader is asked to place special emphasis on the word-modifier 'may', because sometimes the facade of a conflict is misleading.

There have been numerous incidents where I live, all recent, where a victim was beaten to death by his antagonists. On the surface, these incidents may have seemed to be a #2 conflict. No deadly weapons used, just guys and gals out for a good time. In reality, however, they were #3 conflicts. Death occurred. For this reason it would be good to look at a quote from an expert on war:

> "...The conduct of war branches out in almost all directions and has no definite limits; while any system, any model, has the finite nature of a synthesis. An irreconcilable conflict exists between this type of theory and actual practice." [34]

In war there are 'no definite limits'. When a person is accosted by one or more on the street (or anywhere else), it could be a non-life-threatening conflict, or it could be an incident with 'no definite limits'. Any model set up that doesn't take this problem into account, therefore, could cost serious injury or death to the ones following this model. Please keep this in mind.

Recently in my area a pizza was being delivered, when five or six young men surrounded the car of the delivery driver. He thought they just wanted to steal the pizza, and tried to give it to them. They did want the pizza, but they also wanted some sport. They killed the pizza delivery driver, and ran off laughing.[35]

Sometimes when the tyrant attacks, the public authorities can protect you. But often — very often — you must personally protect *yourself.* For some reason many, many people can't adjust themselves to this fact of life. It just won't register. There's about 25,000 homicides a year in our country. That's 25,000 times every year there probably was no cop available to protect the victim. Add to this figure the two million *other* violent crimes, and it adds up to a very small chance that you will have protection when it's needed. Why can't people understand this? And what has the *result* been because people won't protect themselves?

Our crime rate has soared well over 500% in the past thirty years. The bad guys have gotten the message that many people won't protect themselves. If there was more resistance from the victims, the crime rate would go down. It's mathematical.

In just ten years the criminal population behind bars has more than doubled, and it is estimated that it will double again in the next eight years. They can't build new jails and prisons fast enough. Over 2,000 more murders took place in one year in an American city than our soldiers suffered in a *war* on foreign soil! A seven-year-old in Chicago has seen more gunfights than most professional U.S. *soldiers!*[36]

This creeping tyranny can be singularly restrained, if citizens will merely be realistic and see that there aren't enough cops to go around, that self protection is the answer for crime control. The streets are mean and getting meaner by the day. Our homes and loved ones are less safe each year. Fear should be operating a thousand percent higher than it is — fear is the first element in the 'protection food chain'. It is the only thing that can generate the needed alert, so prospective victims of crime will be prepared against the viper, not after the fact, but before and during the fact. This is the scene. This is where it's at. We must accept reality.

Chapter 3: Understanding the #2 Conflict

Stewart and his friends were out for an evening of fun, and it was Stewart's turn to make everyone laugh. The new sport was to approach someone on the street and deliver a haymaker punch before the unsuspecting victim could take cover. The bigger the haymaker, the bigger the laugh. If the victim were rendered unconscious in one punch, the game was won. A bus stop, waiting in line for a movie, anyplace where there were people would do.

They picked a fellow near a university campus. Stewart asked the student a question, and while the victim was thinking about the answer, he got wasted. The student died.[37] Quite a laugh this. Except for the name, it is a real case. And it is given here to reinforce what was shown above, that the line between the #2 and the #3 conflict is sometimes virtually impossible to discern. We live in dangerous times, on dangerous streets.

Nevertheless, in truth the reader must be told how our system works. And it is this: the response to the #2 conflict, moderate attacks on person or property, must always be proportioned. If the victim has overestimated the nature of the attack, and reacts blindly or without prudence, he or she may face litigation, fine, incarceration. On the other hand, if he *underestimates* the nature of the attack, he could place himself and others in serious jeopardy. How, then, to counsel the reader about #2 conflicts? What advice can I give the reader for the above depicted scene of a 'knockout'? How does one respond proportionately, when a muscular youth — who perhaps pumps iron each day, and works out on a speed bag — amuses himself by taking a windup crack on your cranium? Unless he kills or seriously maims, his crime is likely a misdemeanor, punishable by a fine, community service, or at most a hitch in the county jail.

If our society were in its proper state, someone long ago would have grabbed the little dirtbag by his stacking-swivel (an old Marine Corps term) and given him a 'religious experience'. But, alas, our society is *light years* away from being in its proper state. So, back to the question. What shall I tell you about handling the #2 conflict? The truth is, I don't even know what *I* would do in this situation, for sure! I would do the best I could, and that's *all* I'd do. If I could pull it off, he would be in the hospital for a lengthy spell, and I wouldn't get caught for putting him there — my reaction could be a criminal act. But I might lose, hands down, in spite of my A.A.U. boxing and judo days. I simply don't know the outcome. Nobody does, for sure.

This said, I hope the reader understands. The #2 conflict goes 'like a wind tends'. I can't teach the response to the #2 conflict in this little book. I doubt seriously if I could teach it in a very big book. But I can give you the nature of a #2, and its nature is like that of a jury: Unpredictable.

In a fight, if two people are of more or less equal strength and weight, the conditioned reflexes and endurance usually determine the winner. A fight is no different from many other contests. The outcome depends on forethought, preparation, talent for improvisation, one's action under stress when confronted by physical danger, one's prowess with the so-called 'gross motor skills', one's ability to withstand pain, one's ability to inflict pain, one's endurance (or staying power, as some call it), and, of course, the will to survive. All these components are in the combative soup. And one side has more or less than the other. Who has the most, wins. And, like the S.A.S. motto goes, 'Who dares wins'.

The 19th century war theorist Carl Von Clausewitz said that experience on the battlefield (which is what we're talking about), is like oil, which lubricates the wheels of an otherwise dry machine. *Experience.* This doesn't mean that if you're a martial arts student you'll win a street fight hands down. Your opponent may have been bloodied so many times, and have so much scar tissue, that the fight will barely cause him distress. The experience in question here is often only learned under fire.

In spite of my inability to teach the #2, I would feel amiss if I didn't give some opinions. Moderate attacks on the person of a man, carry a different tenor from those on women and children. A physical attack on a man can result in a simple street fight, what we could call a traditional street fight. There are still lots of them around. A physical attack on women and children, however, is very often an abduction for rape and death. These are not #2 conflicts, but #3. The #3 has the direct potential for incurring great bodily harm or death (covered later on). It is vital to bring this to the reader' s attention.

Resisting the quick fix:

Across our land we see sprouting up a new genre of martial arts studios, which pander to women and children.[38] These are not the traditional institutes for training serious students of sports, such as judo or boxing; or the various martial arts. These take *years* to master. This new line seems to be, rather, along the order of fast foods shops, with the objective of training women and children to fight off adult male attackers, and getting them back on the streets as soon as possible.

A little bit of learning can cause a person to be presumptuous, and a little bit of martial arts training can do the same. This presumptuousness can generate a mindset of not having a healthy fear of one's opponent — and that's big time trouble. A professional boxer has a well-developed fear factor. Watch him in the ring. He has trained and fought for *years* to win his fight, because of this fear factor. Without it he doesn't train and fight. Anyone who receives a modicum of martial arts training, expecting to succesfully face a rapist or other violent offender can be likened to a hopeless visionary who has just read a manual on karate, facing a ninth *dan.*

B

All #3 incidents are bloody serious, and can — and often do — result in death. Therefore, they have to be approached with the mindset of 'Kill or be killed', and the training for a #2 incident doesn't do this[39]. This is not true because I say it; it is common sense.

Resisting serious assaults without the necessary hardware, can be acutely dangerous! The best hardware is probably a gun. They're designed to take out the other guy before he does something hideous to your body. That's their reason for being (at least as far as #3 conflicts are concerned). They were made to keep you alive, instead of being butchered.

This subject is covered nicely in a recent, well-documented, book which should be read by everyone doubting my statement.[40] The author, a professor of criminology, speaks highly of guns to save lives. If females or children believe they are in the possession of enough martial prowess from their classes to engage the attacker, this attitude might get them killed. Rapist-murderers don't often mistake the strength of their quarry, quite the opposite: "A curious trait that psychopaths often have is extrasensory perception about others and the ability to pick out personal vulnerabilities with uncanny accuracy."[41]

So, back to the problem, of not getting the #2 mixed up with the #3. Unrealistic notions about personal strength are hazardous, and women and children who have these notions are courting possible death. This error could be the deciding factor which causes death. They could be trying to make a move, when they are already checkmated. This is exactly what the person familiar with the Formula will avoid.

As for the male species, the crime of forcible violation of a woman or child is in a different light for him. His sex is the *offender.* The male is the *attacker.* Different libido, therefore he is less likely to confuse the #2 with the #3.

Under ordinary circumstances a female's or child's defense in any hand-to-hand struggle falls into a different dimension from the adult male's. The female is operating with not much more than one half the male's upper body strength.[42] And obviously the child has even less. It was recently found that during a six month course for women bodyguards, even after they were well into the course, "the women were still not much of a match for anything but the punching bag."[43] The brains of women simply react differently to aggression.[44] Physically, psychologically, women and kids are different from men.

The child needs protection from adults, and this should begin with insuring he or she doesn't have a false sense of security. Children can be nurtured to have a healthy *fear* of prospective enemies, and to not take the first step in becoming a victim.

As for the female, she has experienced life with the heart of a female. Her drives, her desires, her thought patterns are female, all generated by female things. There are *sixty* 'significant' differences between males and females.[45] All this acts as a great counterpart for the other genre. This is a noble thing. Beautiful. And we shouldn't be afraid to cite this up front. In today's egalitarian society, so paranoid about the existence of prejudice, simple predication is no longer a virtue. But in security it still is. In protection it's facts that counts, not opinions. Facts must be faced squarely if one is to learn the Formula. A female should be oriented to use deadly force if she's fiendishly attacked — a gun. If it's not a #3 she hasn't lost anything by having a gun ready. If she doesn't need the force she can put it away. But if she does need it, she's still whole.

In finishing off the lecture about the #2 (which applies even more to the #3), a simple apologetic is given. I will tell the reader what he or she already knows. That you could take the next several years and become a martial arts expert, and gain a mastery over the #2 incident. But, if you do that, there still is no guarantee about your security. A friend of mine, an off-duty cop at the time of his incident, came home one night at about 1:30 a.m., to find a young man jacking up his neighbor's car. He accosted the suspect. He was a veteran and a professional, was good at hand-to-hand combat, and was armed with a loaded firearm. A scuffle ensued, and he 'Sundied' the kid. The suspect was out cold. But, before he could do anything else, three more kids came out of the bushes. They worked him over with a jack. Then they stole his firearm, and would have killed him with it, had it not been modified for left handers. He was in and out the hospital for months. The department finally had to give the officer a disability retirement because of his injury. You may think you will do better than this professional. And maybe you will. Maybe.

Please remember also that there will be police involvement in many street incidents . Any details you can remember will be helpful to them in identifying the bad guy. Also, remember the following. When any two things meet, as in an attack, or when two cars collide, or in many crimes against property or person, *something gets left behind.* This is called the 'exchange principle', which was developed by a man named Edmond Locard in France in 1910 (who was inspired in his task by none other than Sherlock Holmes). This principle says that "whenever two objects come into contact, there is a transference of material from one to the other".[46]

On a large scale, if two cars crash, there will be some kind of debris that gets dropped at the point of impact (that's how the cops usually know who was wrong), and there often will be evidence of the collision *on* the vehicles, such as paint transfers. But the same thing applies when *any* two things meet. So if you're a crime victim, try and remember there will be evidence of the bad guy. Try and preserve it if you can. Some of the modern capabilities for making a match are astounding. One example is DNA typing (deoxyribonucleic acid), where minuscule body parts or fluids (I'm tempted to say of the pico size, i.e. one trillionth the size of minuscule), which is left behind on the victim, can be matched with the suspect.

Now let's move on to more danger. Recommended reading (and visual aids) for the #2 conflict: for about $80 (includes tax, shipping and handling) you can get a training aid for street fights. The package contains a video with two parts, and a booklet. It's called "Secrets of Modern Professional Warriors...How to Win Any Fight in 4 to 7 Seconds". Write to: Tactical Response Solutions (TRS), 2945 S. Mooney Blvd., Visalia, CA 93277. Or call 1-800-899-8153 (Department XT-05). Caveat: in the measure any methods taught by TRS does not conform to a #2 situation, it should not be used in the #2, but rather for the #3.

Chapter 4: Getting in Shape For Death Struggles

Professor Weaver was big on dialectic (examining opinions or ideas logically). He made this statement, which is necessary to understand before going on with the program. "In dialectic the student will get a training in definition which will compel him to see limitation and contradiction...In effect, he will get training in thinking, whereas the best that he gets now is a vague admonition to think for himself."[47]

He means simply that people in modern society have a tendency not to focus on the real, but rather to view things *subjectively.* The old saying that 'beauty is in the eye of the beholder' expresses the problem.

"Only a few sensory experiences are considered universally and innately beautiful. Among them are the following sights: fields of flowers in bloom, rainbows, waterfalls, young children's smiling faces, the sparkle of precious gems, and the view of the night sky filled with stars."[48]

If one prefers (for its own sake) an old tin can with rotting food remnants and maggots, to a waterfall surrounded with flowers, his thinking is bad. He needs to, 'get training in thinking' as the professor said. So here we go, on a basic two minute course in logical thinking.

Generally there are two ways to think logically. The first is *deduction,* the second is *induction.*

1. Deduction. The kind of thinking which can give *proof* of the argument, if done properly. For example, one can say:

   All ships are for sailing.
   The Mayflower was a ship.
   Therefore, the Mayflower was for sailing.

2. Induction. The kind of thinking which gives *probability* of the argument, if done properly. We use this kind of thinking all the time. Thus, if I have experienced enjoyment from a certain brand of beer in the past, I reason that I will do so in the future. Basically it' s simple analogy. Or there is the kind of inductive logic in the courtroom. By listening to two lawyers present their case, the jurors are led more in one direction, and they thus formulate their verdict.

What's all this to do with the operation of security in #3 conflicts? Quite a bit. What you will be reading for resolving deadly conflicts may be disturbing, but it needs a logical mind to make a good assessment of its truth. The logic of what follows is in keeping with the two minute session given above. I won't set up logical syllogisms and arguments (though this easily could be done, and some readers may choose to do just that). However, it is hoped that when the book is finished, the logic of my thoughts will be consistent with the Formula.

All the Formula principles must be utilized in conflicts which can result in death, but the use of security tools (read 'guns') causes many people to develop a tick. I suppose this is normal, because guns are scary, in a way. But cars can be very scary, too. Most drivers don't see, as a regular part of their work, fatal traffic collisions with heads chopped off, heads split in half, bodies crushed and bodies burning, feet chopped off, legs and arms torn off. But professional officers who do see these things, can see cars as just as scary as guns.

Because cars are part of the everyday life of citizens, because cars do us a daily service, are comfortable and pretty, have stereos and temperature control, they're classified by most people as good, while they see guns as bad. But vehicle accidents cause many more deaths per year than guns[49], and "The data from the 1990 Harvard Medical Practice Study suggest that 150,000 Americans die every year from doctors' negligence, compared with 38,000 gun deaths annually."[50]

For many people guns are simply not familiar. They are, in fact, very alien. They're seen as something only that destroy, which is their purpose. But to destroy can be good, if a greater good comes from that destruction. If I destroy a mad dog which is about to kill my child, that is good. The principle holds if a person is destroyed. So, to be against guns and not cars is prejudicial. If the human element is removed, obviously cars destroy, too. It would have been better for mankind if guns had never been invented (would you believe Benjamin Franklin wanted the *longbow* for America's basic infantry weapon?[51]). But doesn't this hold for the internal combustion engine? Let's face it. We're stuck with both of them.

So it's just a matter of education: guns can be seen as something to *serve*, just as cars. This is the kind of 'training in thinking' the professor was talking about. There is no place in security for sentimentalism, or for pacifism as an ideal. Sentiment, yes. The hope for peace, yes. But these are different in kind from sentimentalism and pacifism. Some minds, however, will object to this.

The incredible results from a recent Roper Organization poll, has revealed that as many as "one out of five Americans could be willfully stupid."[52] That is, they could refuse to face facts even when the facts are self *evident.*

Impossible!
2+2=4

This kind of person will not be influenced by this book. As for the rest, it's up to you. Try not to be relativistic.

Several hundred years ago people used to live according to a principle called 'the principle of contradiction'. Its logic is very sound: something cannot be, and not be, at the same time. It was an anti-relativistic principle. It's of interest to us here, because it means that you can't not protect and protect at the same time. You must make a choice. If there' s no other way to save your life or the life of a loved one, than with water to put out a fire, use the water. If there's no other way to stop a madman or a mad animal from mauling you or your loved ones than a firearm, use it, and be glad for it.

Section B: Preparing for Death Struggles

Chapter 1: Guns Can Be Beautiful

> "The U.S. Bureau of Justice Statistics has repeatedly shown that guns are the most effective means of self-protection."[53]

Now we're at the easiest part about conflicts, because the rules here are much more clear-cut than for the #l or #2 conflicts. Tactics (the method for handling the #l conflict) can run the gamut from Do, Re, Me all the way to Ti. And the #2 conflict must be tempered, because there are variables galore. But the #3 conflict, being a hard conflict, is locked in stone. If someone's going to mess up your body, maybe unto death, there isn't a whole lot of selection. You have to neutralize his capacity to act. Physics all the way. Matter and energy. Force. Enough force to take out the bad guy.

"We couldn't see the body until we got to the dike it had fallen behind, and there I saw something I couldn't believe. The man lay, spread-eagled on his back, and most of the rear and side of his skull was gone. Some of his brains were in his pith helmet, some on the ground, and the rest was slowly oozing from the hole in his head. Yet he was still breathing. I turned to Chuck to tell him the guy was still alive, but he hushed me with a wave of his hand and said, 'Come on, let's get this done.'"[54]

What bullets do to flesh, brains and bones is not supposed to be comely. The above quote is enough to show that. The person who got hit was an enemy soldier, but it could be your friend or child or spouse — or yourself, for that matter. This is why defense must take *tools* and *force* into consideration. The quote will also reinforce Natalie's plight: stressed people, whether they are exposed to gruesome head wounds as a way of life, or suffering from severe family difficulties, sometimes need extra help.

Natalie's plight:

The cigarettes were really helping Natalie. She had quit seven years back. But two months ago a bomb was dropped on her. All in the same week she had a runaway son, a husband laid off, and a daughter in a serious traffic collision. She had found some work cleaning houses, which helped a little with their financial problems. But the energy it took, both mental and physical, to try and keep the lid on their lives, along with the tension she was suffering, made the tobacco seem like a big help. She remembered something her father said about it. He was a World War II veteran, and told her once that sometimes, when lying in a foxhole for extended periods, his cigarettes were as important as his rifle. He explained to her how nicotine stimulates our neurons, by mimicking the action of something called acetylcholine (it's pronounced ass-it-til-coe' leen), which in turn reduces the tension and increases the resolve.[55] "Sometimes, when things are tough, " he explained, "we need some extra help."

Ted (which was her father's name), being a cracker-barrel philosopher, even went further. "Natalie darling," he went on, "don't be afraid to use outside help when you have to. While we have a spiritual side to us, we're also material, and we must use material things. The very rifle I used in combat showed me that. It was like the nicotine, in a way. It caused something to happen, something that would help me stay alive. I'm alive today because of that rifle. And who knows, maybe the nicotine played some role in it. Even though I don't use the nicotine now, I don't condemn it, no more than I condemn guns. I think of it like I think of a gun. When I need it, I use it. When I don't need it, I put it away. But both were my best friends when I was scared and hurting in that foxhole."

Death struggles:

"...The facts of war are often in total opposition to the facts of peace...The efficient commander does not seek to use just enough means, but an excess of means. A military force that is just strong enough to take a position will suffer heavy casualties in doing so; a force vastly superior to the enemy's will do the job without serious loss of men. "[56]

This statement, about combat in the modern age, has a priceless value for us. When attacked by the viper in a #3 conflict, you don't want maybe enough force, you want plenty. If the viper is fire, you select the proper fire extinguisher: Type A for wood and paper (or some water will do), Type B for petroleum offshoots, Type C for electrical. If the viper is a scumbag who's going to open you up and kill you, you *snuff* him.

Guns are by far the best tools for this. Well-engineered guns (no cheap stuff, please, you may well regret it), guns with hefty caliber's, used to direct a bullet at a central torso strike on your assailant, "the chest and abdomen form about fifty percent of the surface of the body presented, when upright, to enemy projectiles; skin surface covering the spine and great vessels — the heart and the major arteries — is less that half of that."[57] You must consider the law on carrying one of these tools.

In many, if not most, states it's a misdemeanor to carry a loaded handgun concealed (check your own state, because there's a recent liberalization of concealment laws in several states). This means a fine if you get caught, or possibly even a thump in the county jail. You must weigh this question carefully, taking into consideration where you live, the law, the risks, your own skills - and these just for openers. Some people think, of course, that their life and limb are on a higher scale than facing a fine or jail sentence. One of these was Bernhard Goetz, a man who possessed the Formula in his bones. In the early afternoon of December 22, 1984, he was accosted by four youths on a New York City subway, who wanted to rob him. A few years earlier he had been mugged, after which he decided to take security measures, in case his life was threatened again. Very few words were spoken during this incident on the subway. Goetz sat next to four youths, and one of them asked, "How are ya?", following up with a request for five dollars.

Goetz asked him what he wanted, after which the youth became more honest as to the foursome's intention. It came in the form of a demand, "Give me five dollars!" With that Goetz opened up with a .38 caliber revolver, dropping all of them. And in the trial that followed, the jury cut him loose on eight separate charges of attempted murder and assault, and four assorted endangerment and gun charges. Great thing juries — most of the time anyway.[58] (Don't fail to read author's warning on this type of incident, under footnote 58. C.).

Packing 'heat':

Many officials, public and private, can obtain concealed weapon permits (or hire legal gun packers) for protection, but this doesn't apply to the average John or Jane, especially if they're in the lower class bracket. In fact, "Of 16 law enforcement departments surveyed, not one issued a concealed weapons permit to a minority female."[59]

Some states — or, in some cases, regions of a given state — are more just than others. But, by and large, for the mainstream population, there is a virtual phobia in officialdom about sanctioning gun possession on the streets. It's easy to understand, when the fear is for guns to be in the hands of unqualified persons (we should all be genuinely afraid of this). But this isn't the case. It's fear way out of proportion, fear run amok.

I was recently told a story by the police officer who handled the problem of citizens' requests for concealed weapon permits in a major California municipality. The story staggered me. A retired FBI Special Agent applied for such a permit, and he was turned down. Their policy was absolute: nobody gets a concealed weapon permit. Period.

When this trend is considered in the context of our society's crime crisis, it makes no sense. Civilians probably kill more than three times the number of bad guys than cops do.[60] Trained men and women can be a big help to the police, which is to say to the security of the people. Yet in some locations — in many, in fact — they're treated like pariahs. This trend, of forgetfulness for the citizen, and favoritism for a pseudo-elite (irrespective of threats), is actually frightening. In at least one case that has surfaced, a state Senator in California was hiring a personal bodyguard at taxpayers' expense. Why? Because when this elected official had to go to Los Angeles on business, "It is very treacherous there."[61] Kind of gets you, when you consider that in the April, 1992 rioting, L.A. politicians halted ammunition sales and gun pick-ups to the public, which they desperately needed for self protection.[62]

While in good conscience I can't recommend breaking the law, we should all lobby, not only against tampering with Second Amendment rights (private ownership of firearms), but for *extending* them, for general safety. It was recently admitted that as high as 70% of the students in a training institute for gun self-defense courses, teaching mostly females, carry guns without permits.[63] Good citizens should not have to break the law to live safely.

If the law-enforcement community would allow itself to be co-opted into the armed citizen dispute, and be open to discussion on the merits of citizen preparedness, a partial solution to our crime dilemma would begin. But Chiefs of Police and Sheriffs are presently hamstrung, because a juristic-backed press, itself backed by anti-gun currents of hysteria, threaten impeachment of their very careers if they openly recognize what an armed and trained populous can do to the crime crisis. Hence it is the *people*, through legal organizing and lobbying, who must take the initiative. Law enforcement officials need to have the path cleared before they can take the necessary steps. It is laymen (non-cops) who must work with city fathers and policy makers to make changes, and also bear the expense for the

training itself. If successful, a new dimension of force will appear on the horizon, a force that the criminal element will be afraid of. This will not be vigilantism, but something closer to what are now police reserve units — cadres, trained in police lore by retired cops, under the aegis of the Chief of Police or Sheriff. But because of police budgeting problems, they should be self supporting — and through charter, self-liable for wrongdoing. This licit citizen militia, a respected *new order*, will act as ballast to the growing ranks of citizens now carrying loaded and concealed firearms under the new "shall issue" laws. The latter is a positive trend (though one that could develop excesses, hence the need for the *new order*) now effective in numerous states, which makes it mandatory for law enforcement authorities to license citizens to carry concealed weapons.

The basic law is for our protection. But in 1993, 86 percent of the victims of rape, sexual assault, robbery and aggravated assault said their attacker was armed with a handgun.[64] A just authority cannot tell potential victims to face their attacker weaker than the attacker.

Owning enough 'heat':

The American citizenry is being given a suede-shoe salesman's con job of utter nonsense about guns, by both political leaders and the media. Those who are oblivious to citizens' security, are harping these days about firearms taking a clip of more than the standard quantity of rounds (about seven). 'Assault rifle' is another knee-jerk word, but the fever is against any weapon, long gun or hand gun, which is fed such a clip. These are all trick shots. The issue involves semi-automatic firearms, which is the kind of gun that shoots a round every time the trigger is squeezed. But any 19th century self-loader has this capacity, and in *principle* so do 19th century double-action revolvers (for purposes here discussed, the only difference between semi-autos and uncocked double-action revolvers is in the trigger-squeeze, i.e. pressure coefficient, but this is an issue of *degrees*, not substance). Both of these vintage guns are still around, in households and on the streets of America.

I have a World War I vintage double-action revolver, very serviceable, which I can speed load, using half-moon clips, and get off twelve well-placed rounds in ten seconds or less. So if the people fall victim to the scare of 'long clips' today, tomorrow they fall victim to even the relics. When principle isn't the deciding factor, and emotion becomes its surrogate, you lose.

This country is loaded with guns, over 200 million, and it's my bet that the majority are the kind that will spit out a shot with each click. These fast talkers are capitalizing on the wrong kind of fear, the fear that makes people cower. In some states their pitch has already succeeded, and guns with long clips are no longer allowed for the citizens. But you can bet the bad guys will get them. Because they either steal them[65], or go to states or even countries where they can get them. The mathematics of this circus act is not hard to calculate. The reprobates are getting more fire power than dues-paying citizens. Bad deal, this, and it may get worse, unless the truth gets known.

In support of a California measure to restrict long clips, the Peace Officers Research Association of California wrote: "The military and law enforcement have a demonstrated need for these combat-oriented magazines, the civilian population, in general, do not."[66] What kind of statement is this? America's people are averaging close to 25,000 homicides a year (not to mention the other two million violent crimes per year), and the vast majority are perpetrated against *civilians*! The above statement, that civilians don't have the need, is false on its face. Law enforcement line-of-duty deaths have been about 30,000 in the last two hundred years, [67] five thousand more than for a *single year* in the private sector. The private sector *does* need firepower.

It's a mockery of social justice, just like not letting the citizens buy ammo, or pick up the guns they bought during the L.A. riots. Our country has been through this quackery before. In 1920, the 18th Amendment was adopted, authorizing Congress to legislate federal prohibition of manufacture, transportation or sale of all beverages for intoxicating purposes. The resulting flagrant violations led to gangsterism, racketeering, bootlegging and carnage. This created such a wave of public protests that it was later repealed, in 1933. But in between those thirteen years, crime and violence raged.

Utopianists always foul up the works. In the 19th century they once banned the *Bowie* knife in some of the Southern sates! (which did *nothing* to diminish its popularity, but rather the opposite[68]). This futility of federal big-brotherism was even cited by Dr. James Todd, the executive VP of the American Medical Association, in deciding *against* calling for a ban on pistols.[69]

To repeat, we should lobby for balanced gun laws, and this lobbying should be for both the sale and street possession of handguns and longuns needed for protection, so the poor and the weak in our society can attain security. This should include 'long clip' weapons. The issue of 'assault rifles' is phony, because they were used in less than 1% of all homicides, and as Abraham N. Tennebaum pointed out, in Washington, D.C. no citizen may even *own* a gun, and they "desperately need a way to defend themselves."[70] To prevent gun ownership is an Orwellian horror. It is plainly obscene, and nothing less than wanton tyranny.

During the 1992 L.A. rioting, it was guns that saved the shops of the little business owner, guns and the willingness to protect.[71] This is supposed to be a democracy we live in, so the rank and file should be able to influence laws that can save their lives, and the lives of loved ones.

Pleading for protection:

Surely anyone of good will and sound mind can see the current dilemma in the United States. It would take five *million* additional police in the U.S. even to get our crime rate back to the 1960's,[72] and this is not going to happen. Citizens must be legally armed to meet the dangers theyt face. Our lives, in truth, are vastly more important than any despotic laws that would put them at risk.

Some people can afford bodyguards and other high class security, and some live in places less dangerous. But most don't. Most need to take the Formula very seriously, because they can't hire someone to do it for them. This necessarily means having the proper tool for protection, at the *time* of danger. To repeat my own sales pitch, this is not another way of saying 'read my lips, pack an iron'. Rather it is a clarion call, and a challenge for the citizen to get involved.

Many, if not most, of our public officials, in the deepest recesses of their minds and hearts, would rather see the bad guy get blasted and the innocent victim unharmed. But they need their constituencies to clamor for this to happen, at a time when there's no other viable answer except a police state, a monstrosity our Founding Fathers sought to prevent by drafting the Second Amendment. Current federal law defines this to mean the body of *individual* citizens[73], in spite of what distorters of American history (such as Handgun Control Inc.) allege.[74] Even federal government spending, the bane of so many taxpayers, supports civilian shooters by providing "40 million rounds of ammunition and other supplies annually."[75]

I don't know what one or the other public figure might say, after the fact, about a violation of a gun statute that saved a person's life. I'll leave that to them. But I can't imagine a lawmaker or anyone else saying before the fact, 'If someone's about to place your life in mortal danger, don't be prepared to save it.' The estimates about guns thwarting criminal acts is impressive. Some examples:

A. Over 900 people a day use a firearm to protect themselves in this country[76]

B. 650,000 American citizens use a gun every year to 'thwart criminal assaults'.[77]

C. There are '2.4 million defensive uses of guns of some kind' every year[78]

D. In Kennesaw, Georgia, after a law made it mandatory for every household to have a gun, the crime rate dropped by 75-85%.[79]

E. After the handgun ban in Washington, D.C. the homicide rate tripled.[80]

F. After Florida enacted its concealed weapon law, the homicide rate dropped 10% in Miami, 18% in Jacksonville, and 20% in Tampa.[81]

G. Following Florida's 1987 "shall issue" concealed weapon law, nineteen others states have followed suit, and the states with liberal concealed carry laws have 22% less violent crime than those with severe restrictions.[82]

H. "In self-defense, or for some other legally justified cause, citizens shoot and kill between 1,500 and 2,800 felons annually; this is 2.5 to seven times the number of criminals slain by the police."[83]

I. There are "about 8,700-16,000 non-fatal, legally permissible woundings of criminals by gun-armed citizens" each year.[84]

J. "According to research conducted by James D. Wright and Peter H. Rossi, the barbarians prefer the weak and defenseless. In a survey of 1,874 felons, some 40% of those responding said they had canceled plans to commit a crime because they feared the intended victim was armed. More than a third of the respondents had been driven off, shot at, wounded or captured by an armed victim."[85]

K. "....armed citizens are responsible for more failed felonious enterprises than police officers ever will be." And, "....the majority of rank-and-file officers tend to favor the rights of the armed citizen."[86]

All this should be considered by the anti-gun segments. Cicero said that "The welfare of the people is the supreme law". When the law becomes a procedural surrogate for common sense in matters of life and death, the welfare of the people goes down the drain. It depends on being prepared for the worst, *at the time* the worst happens.

Only 15.1% of violent crimes occur in a person's home.[87] To be secure, one doesn't forget about the other 85%. It would be like saying you're only going to put out threatening fires 15% of the time.

If you're walking in terrain where there's coral snakes you wear high leather boots 100% of the time. It's common sense. Yet many people don't get the point, and the 'coral snakes' on the street know it. They know that a lot of people won't wear these boots at all, on the streets or at home. And they know the chances of a police officer being there to protect you are nil, especially if they time their strike. And the police know this as well, even though they would like to be there. Down deep the police want your security, they risk their lives for it.

Some progressive police agencies across the land are constructing programs based on 'citizen involvement', and signals are flashing that public and private sector law enforcement are merging.[88] This is good, this is excellent. One progressive leader in the law enforcement community sends out 700 armed civilians at a time to help keep the peace.[89] And in the swank Upper East Side in New York, 500 armed security guards are being sought to augment the police.[90] The more this trend continues, the less paranoia by officials about guns. This is the thrust of my whole message, for the citizens to get involved and provide for their own security, the security of their family, their business. Vegetius of Rome said that, "If you want peace, prepare for war". Good saying that. Very relevant to our crime problem in America.

[Author's note. I didn't cover what are sometimes called 'active deterrents' for controlling violence, such as tear gas sprays, stun guns and the like, simply because they are border-line controls. They are used for controlling a violent person when that person isn't armed. In prison and police work they have their place, but they also have their controversy. In the incident which sparked the 1992 L.A. riots, Rodney King got hit twice with a professional stun gun, and still continued the fight. That's 50,000 jolts of electrical energy times two. I want better protection than that if my life's on the line. And if you're thinking about using Mace (a trade name for the chemical agent CN), you may want to think again, because it sometimes makes the attacker even more violent. Sprays that contain 0C (made from the oil of the pepper *oleoresin capsicum*), may work a little better[91], but are "only a distraction", and do not incapacitate.[92] Check your state's laws on private use. There are also 'passive deterrents' on the market, such as hand-held squealers. There are probably close to a dozen brands to choose from. They put out around 100 decibels of sound, or more, which means the sound can carry up to half a mile. The cost hovers around the price of a bottle of moderately priced wine. Check the yellow pages of your phone book for security stores, or gun shops, that often carry protection sprays and squealers.]

But please don't miss the point. In preparing for a life and death struggle, you cannot assume that your opponent is not going to be armed. You want enough force to stop the attacker. You don't want to *maybe* scare him, or *maybe* have enough force. You want to save your life, or the lives of loved ones. About one in a hundred rapists, for instance, kill their victims[93] If you're one of the chosen victims, do you want to assume you won't be the one?

Water (if sprayed, not streamed) may put out a grease fire, but it's not like having a Type B fire extinguisher. By the same token, guns are by far the best tool for saving your life in a death struggle. Don't listen to people who tell you the death penalty isn't a deterrent. If you shoot and kill a guy about to do you in, it's a deterrent because he's not going back to the streets again. He's deterred forevermore.

Chapter 2: Feeling Fine About Protection

When danger strikes:

Vivian lives with her husband and children in a somewhat remote area of a county with a low population. She's been to a school-board meeting in town, and is now traveling home. A car is following her. While getting into her car after the meeting, she thought she saw a man studying her. And she's now worried that this is the person following her. On a rural, county road about four miles from her residence, Veev (as her husband calls her) has a flat tire. The nearest house is over a half mile away. As she limps her car over to the shoulder, she notices the car pulling in behind her own vehicle. She's now downright frightened, and heading swiftly towards panic. A serial killer has been operating in the general area, with the news becoming more macabre with each of his kills.

Vivian's husband often carries a concealed weapon, the local sheriff having given him a permit to do so. And he sometimes keeps the weapon in a specially crafted container under the driver's seat, a steel box where the front drops down via a control switch under the mat, near the accelerator. She prays that the gun is there. Her right foot hits the switch, the steel front drops, and the gun is there. It's a .45 caliber semi-auto pistol, loaded, ready to go. Her husband trained her to use the weapon, and confidence begins coming back as she grasps the checkered grip.

She grabs a high-powered flashlight, always kept in the car, and exits the vehicle, placing the gun in her overcoat pocket as she does so. The man is now approaching her, and is about twenty feet away.

She remembers what her husband taught her to do, if ever in such a situation. He told her to think in terms of noise, heat and light. First she uses the light, shining the beam of the flashlight on the man in a way to see his hands and general appearance. Then she uses noise. In a loud voice she challenges, "What do you want?" He answers that he wants to help her change the flat.

Almost screeching, she states, "I don't need help. Begone! Now!" He cajoles, saying he just wants to be of service. He advances. The 'heat' is removed from her pocket, and she holds the muzzle down, pointed at the ground between the man and herself. But she makes sure he sees the gun.

Quickly the man retreats, gets into his car, and begins to pull away. Vivian memorizes the license plate number. The man is later picked up, and he becomes a prime suspect as the serial killer. It is later determined that he is, in fact, the serial killer. (This is based on a real case, though some of the details have been re-worked.)

True Love

Checking out the surface:

Vivian didn't know the man following her was a serial killer. But she possessed the foundation stone for all security: fear, rightly understood. And she utilized all relevant points of the Formula. The number of the victims of the killer in question (from which this case is taken), were legion. Vivian was the exception. She feared, and from that fear came an alerted and prepared state, with the proper *tool.* But there's more to be garnered from Vivian's incident than this.

In any incident involving the need for protection, a series of surface events take place. Some people will interpret these events in one way, others in another way, and some won't even pay attention to them. But it is the surface event itself, taken one by one, and then as a whole, to which the authorities will turn, when deciding whether you either did right, or wrong.

In Vivian's case, the first element was the fact a serial killer was working. The second was the man studying her. The third is that she was followed. The fourth is that the car pulled in behind her. The fifth is that he kept coming, even after she told him to get away. The sixth would have been Vivian leveling the gun at the man. The seventh would have been a final warning. The eighth would have been the explosion. (The reader should keep in mind that some bad guys wear bullet proof vests these days, but a serial killer acting on impulse for his next kill, might not be wearing one).

When taken with the previous discussion on logical thinking, this series of surface events becomes the building blocks for a story. It is a story of inductive logic. It tells of a probability. Placed one upon the other, what is the probability this series of surface events spells danger?

What would a reasonable person conclude from this story? Our minds are for deciding what things mean. We have no crystal ball to tell, for sure. We must work on placing certain events in a context, and making a decision as to what that context means, in terms of our safety.

In a shorter version, what would the surface events mean if you found a stranger in your bedroom at 3 a.m.? The answer means for me that your safety is 'in deep dumbo'. That' s exactly what Vivian thought about her's. She put two and two together and it equaled a serious threat to her well-being. Then she followed the first principle of security, she defended. When one side is defending, the other side is offending; there's no way out of this rubric. The problems that arise are problems of prudence. And Vivian was prudent. She acted as any reasonable person would have, which is the standard of judgment. In any conflict there is a choice. To engage, or not to engage. There are rules for engagement. All professional conflict resolvers have these rules. The key question in the decision is, "What other reasonable choices did you have?"

If she didn't shoot, she either gave up her weapon, or turned and ran. If she turned and ran, the indecision and confusion could place the gun in the criminal' s hands. (If the reader thinks those are prudent selections, he or she is having a problem with all this. Back to professor Weaver. Back to dialectic square one. But don't give up, you'll get there.)

Section C: Checking Out the Landscape

Placing our dangers in context:

Before we go further, it will be good to view our three conflicts in their natural habitat. Any subject matter is more easily understood, when it is seen in relation to the whole. A doctor doesn't learn about the heart or the lungs as isolated from the whole. He studies them vis-a-vis the circulatory system and the respiratory system, and these in relation to the other body systems. And most importantly, all the body systems are finally seen as they interact, and are integrated as the whole of the human body.

And so it is with our three conflicts. They also interact and are integrated, in a way that can be compared to the doctor's conflicts. A simple malady goes unchecked in one part of the body, then metastasizes and becomes infectious in another part, and felt throughout the whole body. A treatable laceration becomes infectious, and spreads until it becomes gangrenes. It effects the whole body. A microbe in the diet causes dehydration and anemia, affecting the whole body.

When the soft conflict ruptures, it becomes either a tempered or a hard conflict to resolve. And when this hard conflict spreads as a kind of sickness to the whole, it becomes a social problem. This, of course, is where you and I are today. We cannot fix it. But we can define the nature of the problem, and if our definition is realistic, we can take steps for its resolution as a whole. The first step is in understanding the landscape as it really exists. Then we can offer solutions.

A funny thing happened...

When I read the newspapers a funny thing happened to me. I thought I'd heard it all before. Our former chief law enforcement officer in the United States, at a luncheon, said that the answer to our 'epidemic of violence' is for repeat offenders to be given heavier prison terms. Local law enforcement officers and prosecutors objected, saying that that is not enough. They said we also must deal with the 'root causes' of crime, and this would include more gun control.[94]

The only problem with the dictum of these good men and women is that the part about heavier sentencing and 'root causes' is old hat, and the part about gun control is wrong. We can't give hard time to repeat offenders unless the system is changed. There isn't enough room to keep the prisoners, it's that simple. The prison system in America is completely hypertrophied. It' s swollen up. The courts can't give the bad guys enough time, because the system must continually exude to make room for the never-ending input of new convicts. And when any system keeps swelling, it eventually implodes, like the April, 1993 prison disaster in Ohio. Discharge, implode. Discharge, implode. That's the nature of our correctional system in America. So any idea about giving more time, harder time, is a fallacy of the most basic kind, unless the system itself is changed.

What kind of a system change am I talking about? From one that doesn't work to one that works. Our present one is not working and everyone knows it. The Marine Corps brig system, on the other hand, really works. They make it truly miserable for the prisoner; they make his time *hard*, and 'cruel and unusual punishment' accusations haven't stopped it.

So it's a simple problem of math. Make our system work by making the equation make sense. Change our system to resemble the Marine Corps brig system. Not the quick-fix 'boot camp' approach but the *brig* approach. The Marine Corps boot camp system and the Marine Corps brig system are two completely separate entities. And elaborate and enormously expensive prison complexes are not necessary. Our correctional fathers are relying too much on concrete, steel and technology, and relying less on the human factor. One cell in a modern prison can cost the taxpayers up to $75,000 a year.[95] *One cell.*

There are millions of acres of property in this country where decent camps could be built, with land mines to keep the convicts from escaping. In California, the Department of Corrections is building electric fences, capable of killing humans (though presently killing birds which land thereupon[96]), to keep convicts in. The cost is $28 million[97], but land mines would do the same at a mere fraction of the cost. Methods of cost saving are plenty, and this cost saving can translate into prison room for everyone who wants to go, if the right people will merely look at solutions.

The job of corrections should be done by the private sector. In fact, the private sector could create and run prisons and make a profit. They're already doing it in the United States, with up to 40% cost savings, and probably more efficiency than the public sector.[98] And in the process of this private sector engagement, an army of effective, professional prison officers could be created (and also transferred from the public sector) putting men and women to work, at a time when jobs are badly needed.

With this new start, repeat offenders could be kept where they belong, at a cost the taxpayers can afford. The old prison system would be the perfect jumping-off place, the perfect staging camps, for holding and releasing the convicts to the new camps as they are built. A deal could be struck between the public sector, which now holds the keys, and the private sector, similar to the arrangement between Detroit's big three auto makers and the national defense laboratories.[99] Only, in the correctional agreement, the private sector eventually would take over completely, relieving the federal and state governments of a task they're not doing well, because they've simply taken on too much business for themselves. It's time government got out of the prison business, and let nature take its course.

A prison expert recently made an interesting statement. In the laws of nature, sayeth this expert, there "are no rewards and no retributions, only consequences."[100] Therefore, the expert continues, let the prisons not punish, nor let out the convicts for good behavior. We could just keep prisoners incarcerated for separation's sake. Let them be separated from people who are trying to make our society work. Let them work in something constructive that will help society. And let them do *all* of their time.

The current prison officers could roll-over to the private sector during the process. And the utopian institutions (like the California state prison at Pelican Bay, called "barren and automated...an Orwellian horror", and a "computerized fortress"), [101] eventually would wither, and die. (Or they could be sold by the state, and used as exotic hotels). Because human beings can do the job of corrections, a whole lot better than technology.

The key to the new system would be *people*. Lots of people, and lots of guns, and lots of balls and chains. And less technology. Much less technology. For once people should put technology out of business instead of technology putting people out of business.

Unmasking the 'root causes' trick:

As to 'root causes', the superclerks in this country have poured over two trillion dollars into these 'causes' since 1965 [102], and the country is sicker than ever: "Poverty programs actually worsened the problems of the poor".[103]

It's sicker because the money's being pumped into trying to cure the wrong causes. During those thirty years, as mentioned previously, our crime rate has gone up 560% [104]. The problem can't be fixed just by dumping money into 'programs'. Our leaders must recognize the real problem, which is the disintegration of the family. This is the real 'root cause' of our crime and violence in America. An interesting study was done, over thirty years ago, by a veteran judge in the criminal court system. He was so disgusted by crime among the youth in this country that he began studying western countries that had low crime rates among the youth.

The lowest, by far, was Italy; so the judge traveled there to see why. And his conclusion was not only because they had strong families, but because in those families the father was the head of the household. The title of the article by this judge was, "Nine Words That Can Stop Juvenile Delinquency", and the nine words were "Put father back at the head of the household".[105] Judge Leibowitz's conclusion was confirmed by a 1988 study, which found that *predictions* could be made about violent criminals based on this one point: no father at home.[106] Eighty percent of American kids come home from school with nobody home at *all*. Why should anyone be surprised by our crime rate?[107]

If the fact of disintegrating family life is not faced, no amount of money or 'programs' is going to fix this problem. And there are ways to fix it. It should begin with the Executive Office itself, the White House, where the sovereign of our land uses his muscle to attack the problem. He should take a stand against the filth and violence coming into homes on TV,

and in the movies. He should campaign for legislation and other sanctions to keep fathers home, laws with real teeth in them, laws that make the fathers comply (and mothers, in many cases) or go to the new prison system, with plenty of room, and plenty of prison officers, to do hard time. And review also is *badly* needed of divorce laws that arbitrarily shaft the father.

What I'm suggesting here is not a law simply to make 'deadbeat dads' support their kids, such as the state of Massachusetts has[108], but a law that makes men *afraid* to go around acting like stray dogs. What defense can a man possibly have that he should beget children and not support them and their mother? Even *Hollywood* used to support responsibility for fathers![109] We have to start somewhere to build the real infrastructure of this country, and it's to look at what the main infrastructure is. The family.

Focusing on first principles:

As for gun control, up to 70% of the guns used in crimes are stolen [110], so what's more gun control going to do? Unless outright confiscation of privately owned firearms is effected, (which would precipitate the next American civil war), more controls would be counterproductive. Radical gun control means a police state, and after that, totalitarianism.

Crimes must be stopped as they are happening, or more precisely, a split second *before* they happen. Infectious epidemics need inoculation, so what needs to be fed into the diseased area is the 'serum' that will work: a bullet in the body of the person about to rape or kill or mutilate.

People are unsafe and getting moreso each day. But *real* solutions are needed to resolve the problem, not demented laws such as three-strikes, which is going to get cops killed[111], and which give life terms for a $5 cocaine sale[112], or a bike theft[113]; or equally demented federal drug sentencing guidelines that even *judges* refuse to accommodate[114]. Besides being unjust, these laws floods our prisons with even more people[115], making them too full for serious offenders. It is fanatical legislation, a desperate grasp for correcting a problem without good analysis, and it is right out of a Victor Hugo novel, the slave galley for stealing bread.[116] And the cost! California will have to build twenty more prisons for three strikes (at a cost of $500 million each), making forty eight prisons in that state[117], and create bureaucratic nightmares of court backlogging.

When a violent criminal commits a crime, he does it when the victim is unprotected, in a spot he has chosen to make his violence and exit safe for himself, and hazardous for his victim. *This* is the first principle we must deal with in crime and violence control.

A great teacher once stated that, "The staggering number of facts to which (modern man) has access serves only to draw him away from consideration of first principles, so that his orientation becomes peripheral". [118] When danger is upon us, it must be dealt with forthwith. When a woman is grabbed by a man twice her strength, and dragged helplessly into a lair, heavy prison terms and 'root causes' are absolutely meaningless to her. The only thing that means anything to her is to stop the action — now — of the person about to defile and maybe kill her. The next day, when authorities look at this incident from the boundary (from the *peripheral)*, and once again say that the criminal who did that dirty deed should get a heavy sentence, and crimes should be dealt with at the 'root causes', are pouring useless salve on an old and painful wound.

The guy should be dropped, *before* he acts on the helpless woman. And the only one who can usually do that is the victim, probably with a gun. That's the first principle that must by understood, and dealt with, if this country is going to be inoculated against violence.

Am I recommending vigilantism? Absolutely not. I'm recommending security. When a grease fire breaks out in your kitchen, the fire department may save your house from burning, and may save the lives of your loved ones. But is that what you want? Maybe? Why not have a Type B fire extinguisher handy so you can put it out yourself, before the fire defiles your property and maybe kills your family? A simple first principle is being dealt with here; not theory, not sophistry. And the same applies to all hazards, all threats, all tyrants, be they fire, water, earth or man.

Section D: Releasing the Energy and Matter

Chapter 1: Practicing To Use Force

Listening to Fast Eddy:

In the movie classic, "The Hustler", Fast Eddy, the pool hustler from Oakland, California was pondering his thoughts outloud to Sarah, about people in possession of certain skills. While musing about his own talents on the pool table, he extended this subject by saying, 'Anything can be great, bricklaying can be great; if (the bricklayer) *knows.*'[119] Fast Eddy made his living, at least in part, by winning from people who *didn't* know what they were doing. Hustlers of all kinds are experts on such things. Whatever the particular bunco game, be it a big con or a short con, a sweetheart con, pyramid con or snake oil con, the hustler has a sixth sense for picking his mark. And often he has a sincere respect for everyone, who 'knows', like Fast Eddy did. What he meant by a person who 'knows', is that the talent or skill is in his bones, in his chemistry. It is a person who can always pull off what he's about, because he knows *ahead* of time he can do it.

It all comes down to 'when to let it go, and how much', as Fast Eddy explained. Timing and judgment. I've also heard it called 'tribal knowledge'. A surgeon has it with his scalpel, a barber with his clippers, a teacher with her student. A veteran cop has it with the subject of his quarry, and a pro burglar has it with his 'target'. And it can be acquired, and become an art, in conflict resolution. There is a type of prescience needed for this trait, something we can call a sense within, something that calls the shot before it happens (the late pool champ Willie Mosconi once sunk 526 straight balls in a row). And there is a technique where it can be learned.

The central element in this prescience, this other sense, is to really *know.* You don't make assumptions, you *know.* In the case of conflict resolution, it is about the threat which you are facing. You don't make *assumptions* about it. You must see reality, soberly, in all its fearfulness and passion, however grim — or in all its benignity. You can't suppose something is threatening. You can suspect it might be, but that's different from supposing. You must see things clearly. The evidence must be there.

Approaching the intersection:

Picture yourself at a blind intersection, a four way intersection, without signal controls, without stop signs, without warning devices of any kind. Trees, signboards, high fences block the view. As you approach this intersection in your car, you must slow way down, or stop altogether. You must listen. You must be keenly alert. Observe the way a cat is when it's watching its prey. All its senses and cells are operating for the moment of the dash. You must sense another vehicle approaching. The other vehicle is your opponent.

Be careful that you don't become overconfident about discerning your opponent's intentions, however. What you're interested in is what he's going to do, not what he may or may not be *thinking*. Is there a car coming or not? *That's* what you need to know. Is your opponent going to strike? What gives you data on this question? What are the surrounding circumstances? What are the noises? The movement? The verification that there is hostility?

Are you alone? Could he be showing off? Could he be just an intimidating type of person, and you are over-reacting? Is there something he wants from you? Is it known that you are carrying a valuable? Are you physically attractive to whoever's there? Is the circumstance of time and place such that you are vulnerable? Vulnerable to death? What makes you think so?

Refusing to leave danger alone:

David Mann, mild mannered and with deep phlegmatic characteristics, in his orange Plymouth sedan, is traveling to pay a visit to a major account in a remote, semi-desert area of southern California. Almost no cars at all are on the road he is traveling, a two lane, old state highway.

The driver of a 5-axle tanker rig has been giving him a bad time. The truck driver's harassment, which at first seemed just the antics of a bad temper on a hot road, have degenerated into serious highway badgering. In fact the last incident was outright scary. The teamster had waved David around, to pass his big-rig, while going up a hill, and a car was coming down the hill at break-neck speed. David jerked his car back into his lane, but the two cars came very close to a head-on collision. Could that have been on purpose?

But now there was no doubt. While waiting for a freight train to pass, the rig moved in behind David, touched the Plymouth's rear bumper, and began pushing it into the train! The caboose whizzed by only a second before the front of the Plymouth was pushed up, onto the tracks. He was trying to *kill* David!

But why, David asked himself? What's the cause of all this? David, now frenzied with paranoia, had tried to make mental computations as to the truck driver's *intensions*; how he could reason with him, and various other assumptions about the attempted murder, all of them irrelevant and immaterial. He didn't try to think of a defensive strategy, or defensive tactics. Nor does he do the most logical thing. Leave it alone, and go home.

Instead he keeps going, until the tractor-trailer combination gets him cornered. It chases the Plymouth off the state highway, onto an un-maintained county road, then onto a dirt road. David spins out and crashes into a bank, and at the last minute jumps out of his car, just as the train-like rig crashes into the Plymouth. He lives, but only by a whisper.[120] He is lucky, because he did the exact opposite of the Blind Intersection practice. He went out into that intersection when he heard another vehicle coming, right into its path. He went into enemy territory. And it almost cost him his life.

The concrete oasis:

Silvia was standing on a concrete oasis, one of those little islands you see on a busy boulevard, placed directly in the middle of the street. Traffic's zooming by in both directions. The height of her little sanctuary is about the same as the sidewalk curbing on both sides of the street. She is safe here until the signals change, at which time she will finish crossing the street to safety. She knows (only too well) that at any time a driver in one of the cars going by, could let the steering wheel drop just an inch or two, and the car would become a killing machine. It would deviate from its path, and come ramming into the curbing on her concrete haven, with the car jumping up onto the slab, taking her out. Her mind wanders back to that horrifying night.

It was less than a year before. She had returned from a date, and was getting dressed for bed, when the doorbell rang. She thought it was probably Rhonda next door, since it was only 9 p.m. She went to the door, but left the night-chain fastened. Looking through the three inch space, she saw a young man. He asked her for directions. Before she hardly began speaking he shouldered the door, easily tearing the night-chain from its axis.

Silvia was no sitting duck (except for her lousy door security[121]). In a fraction of a second she knew. She wheeled around, ran into her bedroom and grabbed her gun. At bedtime she always took the gun from the locked case, and placed it between her mattress and the box spring. The scumbucket didn't have a chance. She had the gun pointed at his midsection a moment after he came through the door at her. She fired. The first slug didn't drop him, but he was probably mortally wounded. She fired a second time right before he lunged, jumping out of the way immediately after squeezing the trigger.

The dead guy was identified very shortly after the police arrived. He had done two terms in the state prison for rape. He had only been on the streets for eleven days when this happened. The officer who handled the case told Silvia, "For the record, I hope my own daughter would react like you did".

The light having changed, Silvia stepped off her place of refuge into the crosswalk. "When they come into your territory", she mused "be as ready as you can. It can happen in a split second".

Chapter 2: Outlining

When speaking in front of people, whether in a courtroom, a high-school speech class, as a public figure, or for any purpose, you must go into the speech prepared. You have to be programmed ahead of time. You can't do it when you get up to speak; at least, this is true for most of us. The hardest way to do a good job is to have everything written down. Just a simple *outline* (with notes on specifics if needed) is best. Here's an example:

General subject: terrorism

A. Modern international terrorism

1. 1967: The Six Day War

a. Tension mounts between Israel and Arab neighbors

b. Gulf of Agaba closed to Israel shipping

c . Humiliating defeat of Arab forces

d. Israeli troops occupy Sinai Peninsula, Gaza Strip, Golan Heights, West Bank

2. Arabs despair of victory in conventional fighting, begin guerrilla/terrorist operations. Moderate leadership of PL0 is discredited, *al Fata* becomes prominent, splinter groups follow [PLFP (Marxist), Abu Nidal (rejectionist front)]

3. Terrorists from Europe, Ireland, North and South America join cause, generally anti-western.

B. Important names in terrorism, yesterday and today

1. Michael Bakunin, socialist, l9th century

2. Serve Nechayev, Jacobin, 19th century, Revolutionary Catechism, provoke government's repression.

3. The Narodnaya Volya, 19th century, above ground/below ground concept

4. Carlos Marighella, Brazilian, 1960's, Mini Manual of Urban Guerrilla Warfare, terrorize until government represses.

5. Ernesto (Che) Guevara, Fidel Castro's companion in arms, the value of charisma, Bolivian peasants

6. Abimael Guzman, Peru, Shining Path (like Khmer Rouge), currently recruiting in Los Angeles, California, under Revolutionary Communist Party (RCP) [122]

7. Mahmud Abouhalima et al, convicted of the February, 1993 bombing project of the World Trade Center in New York, and suspected of other moves against western civilization, in general, and the U.S. government in particular. Ten later convicted of conspiracy.

C. A new generation of terrorists?...citizens against their own government.

1. Japan

A.Time and place: Tokoyo, March 20th, 1995

B. The event: Sarin (nerve gas), a weapon for mass killing, placed on the subway. Twelve killed, 5,500 sickened.

C. The meaning: anti-government cult acts out. Aum Shinrikyo cult, led by Shoko Asahara, seemingly want to start a world war with plans to attack 'government buildings, the Diet and the Imperial Palace.'[123] Chemical, biological and radioactive weapons can, and will, get into the hands of terrorists[124].

2. United States

A. Time and place: Oklahoma City, 9:02 a.m., April 19, 1995.

B. The event: ANFO (ammonium nitrate and fuel oil) bomb in rented Ryder truck destroys Alfred P. Murrah Federal Building, killing 169 and injuring 500.

C. The meaning: Two anti-government, former Army buddies indicted. Political leaders and the media try to use the tragedy as a partisan *blitzkrieg* against the rights.[125]

3. Israel: Izzedine al-Qassam Brigades, the military wing of the outlawed Hamas (the Islamic Resistance Movement) gains deadly momentum in quest for Islamic Palestine.

Being ready:

When danger strikes it can strike without any warning whatsoever. Like Silvia, you have to have your outline ready. It has to be a part of you. You won't have time to deliberate. So here's your grand outline for handling the Three Conflicts.

The grand outline:

A. Soft Conflict.......silence

B. Tempered Conflict....only the first

C. Hard Conflict........kill

Grand outline analyzed:

A. Soft Conflicts. In noise and trivia conflicts, the hardest part is to control yourself. Your first impulse should therefore be to remain silent. This does not mean you hide in a corner. It merely means you want to step back and look over the situation. In #1 conflicts you have time to do this, to think about problem solving. Just remember that you may be up against a no-win situation. Your opponent may be a master at heckling.

Once there was a heckler at baseball games so good at rattling visiting batters, he was actually hired by big league clubs. He was paid to sit in the stands and menace the opposition with his 'scientific heckling'. His name was Pete Adelis and he was literally a professional heckler. He *won* ball games! One day a batter got so infuriated (Dodger batter Billy Herman) that Herman threw down his bat, leapt the fence into the stands, and chased Adelis, intent on doing him in.[126] Don't get sucked into tangling in Soft Conflicts, except by your wits. Use tactics, like a street cop. The resolution motif of the #1 conflict is always tactics (as opposed to physics). It is hard to imagine a given profession with more expertise in this line than a street cop, since the #1 conflict makes up a huge portion of his or her day-to-day duties. Prison officers and school teachers probably run a close second, however they usually have institutional controls to help them manage noisome antagonists — a luxury that street cops don't have.

There may be disagreement on this point. I've known cops who believe there's only *one* way to handle a puke, and that's to 'take no lip'. Wrong. This standard locks you into what sometimes is a no-win situation, making the cop look bad, because he loses his temper and the puke walks from the scene. Anyway, once your temper is out of control, it becomes increasingly difficult to remain rational. Discern what I'm saying here. It's always wrong for a person in authority to let someone 'walk all over him'. Because of the extended variety, however, of a given #1 conflict, there is no other *standard* for resolution except tactics. This lets the officer, or the civilian, choose from a myriad of responses, giving him more chance at victory.

B. Tempered Conflicts. Here you have less time to think than in the Soft Conflicts, but more time than in the Hard Conflicts. In these #2 conflicts you have to decide how to use your force: to engage or not engage the enemy. This is the old 'fight or flight' choice. And the answer depends on *your* abilities vis-a-vis the abilities of your opponent, plus a host of other risks in modern day America. If you're not up to it, get out. If you can't get out of it, do the best you can. If my memory serves, it was the heavyweight champ Archie Moore, who said he always experienced some fear getting into the ring, but he *only felt the first* punch. If you get into a scrap, try and think like Archie, that you'll only feel the first one.

C. Hard Conflicts. If your life and limb are in danger, if all the elements are there, as in the case of Vivian (Chapter 11), and Silvia (Chapter 12), take him out. Kill. Got a better suggestion?

Section E: Primitive Death Struggles

Shakespeare's Hamlet, the Prince of Denmark, uttered one of the most famous lines in all of English literature when he stated, 'To be, or not to be, that is the question'. His poignant soliloquy has to do with existence itself. Is it better to be, or not to be? I had to struggle with this question in regards to this section on 'primitive death struggles'. Is it better to place this into existence, or not to place? Most people don't think about what follows, and therefore don't know it exists. And some, who get disturbed at blunt savagery, will want to skip this section.

When I was in corporate security, I used to buy books offering instruction in bomb making, electronic spying, drug cultivation and preparation, sabotage, private espionage and the like — to acquaint operational and staff managers with a problem. The problem was (and is): what employees, and everyone else, can buy on the open market. Mentor Publications, Lyle Stuart Inc., Paladin Press, The Diamondback Press, Loompanics, to mention a few, publish books that can be used for good or for bad purposes. Should I bring something into existence that can be used for both? Here was my dilemma, as to whether 'to be, or not to be'.

I opted to print it. My first reason is that the following instruction can, in fact, save a life. The second reason is exactly the same as when I showed the abovementioned books to upper and top management. 'If it isn't written, it isn't known' goes the old saying. If you know what your enemies have access to, it will help you prepare against them.

There's always a 'what if' problem when you contemplate death struggles. What if everything goes wrong? I lose my gun, for instance. I have no weapon to defend myself against being mauled to death. I will lose my life if I don't fight to the death. If the worst happens, what do I do? Since I'll die anyway if I don't fight, shouldn't I do *something*? Since this is an honest book, and the above is an honest question, it should be answered. Sometimes the truth is gory and even painful.

Count your lucky stars:

> "Heavy shelling creates shock waves which numb the senses and cause bleeding from the ears. The danger makes victims' adrenalin run strong; a reaction which in nature is designed to provide the extra energy needed for sharp bursts of life-saving activity. Under shelling, however, there can be no movement, and the victims are forced to lie still under as much cover as they can find, thereby denying the body's demands for violent action to burn out the adrenalin. There is an internal physiological conflict within each man, and he quickly becomes exhausted, or breaks under the strain."[127]

If you ever face death when you are unarmed, and must react violently, thank your lucky stars that you aren't the victim of a shelling, where you wouldn't be able to act at all. You can act without 'internal physiological conflict' in the following situation. You can let go, *really* let go.

Body chemicals:

When danger is upon us, our body produces a hormone called *epinephrine*, which prepares us to fight or flee. A related hormone is also produced, called *nor adrenaline*. These are good, but under certain circumstances, they become a problem, because the fear is generated when we don't want it to. Like in giving a speech .

In any case, the way to control these chemicals is by contracting the rectus abdominis muscle, a long flat muscle extending bilaterally along the front of the abdomen. Further, if there's a feeling of nausea, it's caused by your diaphragm (the breathing muscle, as it were) putting pressure on your stomach. Just place your hands under your rib cage, then with hands still there, take a good breath and bend over, clutching your diaphragm, and stand up again. Repeat as needed. This keeps the diaphragm from being where it shouldn't, and should relieve the bad symptoms.

I used to use the contraction of the *rectus abdominis* before testifying in court, or before making a speech. Actors use it before they walk onto the stage. A close friend of mine, a veteran cop, with twenty-five years on the streets, and a combat veteran of World War II (an Army Ranger, no less), once followed me in making a speech (his speech directly followed mine). I was surprised that he used it. I thought he had no fears. Everybody who knows about it uses it. When I've been involved in situations where my life hung on the next move of my opponent, I don't remember using either of the above methods, because I was too busy on other things. These two tricks are probably most useful when you're waiting for something unpleasant to happen, and conjuring up scary thoughts. But they should be kept in mind.

Bones and stuff:

In killing with your hands, you usually want to focus on the head and throat, but there are other considerations. The adult body has 206 bones (25% of them being in the feet, excellent place to cause pain[128]), and some of them very easy to break. Break as many as you can, especially the big ones, because you must get your prospective killer to go into shock. There's about five or six quarts of blood in an adult. Loss of one quart is serious. Get the blood of your assailant flowing as quickly and as much as possible. Try for an artery, there's two biggies on either side of his throat (carotid arteries), but also under armpits, inside the thighs and kidneys, in the neck, wrists, and inside of the elbows.

There's five levels of skin to go through, so you must get through the first four levels in the *Epidermis*, and go through the *Dermis*, where the nerve endings and blood vessels are. A sharp tool is the best for this, but if worse comes to worst, use your teeth (the front ones, called incisors). Bite, rip, pull out, like a lion.

There are always three aims: stop the breathing (air); get him to bleeding a lot (blood); get him into established shock.

1. Air. What controls all life? Oxygen. What controls this in our body? Motor nerves in the brain do this, in the back of the head, just above the spine. Use something as heavy as possible to crush this part. A very hard blow to the back of the neck also might work, because it might bring about asphyxiation. Slash or tear out the windpipe if you can (blood goes into his lungs when this happens, so death will be from blood loss). Crush his windpipe (choking or striking). Stop the flow of blood through carotid arteries into brain (strangulation). If you go after the lungs, you have to get both of them quickly (one injured lung will only cripple); use a sharp instrument.
2. Blood. The biggee is the heart, get something sharp into it, or at least an artery (mentioned above). Very hard blows to the ribs and/or breastbone can break or splinter them, driving them into the heart, which will get the blood flowing. If something heavy (a strong guy can do it with his fist) is thrust into the floating ribs, they might be forced into his liver, but this might not take him out immediately.
3. Shock. This is often the primary cause of death in traumatic injuries, such as the above, because it involves the nervous system, which is a major system. The body systems are inter-related, so if you get to one in a big way, you will likely cause shock. Spinal damage, vital organs, burning, electrocution, all of these hit the big systems.
4. Blows. Forget the face. We're talking kill or be killed here. Blows to the face are rarely fatal, just watch a fifteen round title fight if you doubt this (but obviously the eyes can be important). In the back, you're after the back and base of the skull, where the neck meets the head, as it were.

Essentially you're after the brain (the central organ in the nervous system). A part of the brain called the *medulla oblongata* sends signals from the spinal cord to the rest of the brain. So you want to stop it functioning by a hard blow to the back and base of the skull. Or a severe blow (preferably with a heavy instrument) to the area between the shoulder blades, breaking the neck and/or deranging the spine and cutting the cord, may serve the purpose.

In the front, if you want to go for the eyes, get a thumb into the eye socket, and press the thumb inward, to the center of the head. While doing this, get the other hand on his head and jerk it hard. Brain damage. In the front, you're looking at the throat area. Here's some terms you should know:

A. Pharynx: common passageway for food and air at back of mouth or nose. At its lower end it divides into two passageways, one for food, the other for air. Food is routed by muscular control in the back of the throat to the

B. Esophagus: the food tube, leading to the stomach (it's about 9 inches long).

C. Trachea: air tube, or windpipe, leading to lungs.

D. Epiglottis: small flap of tissue separating esophagus and trachea. It acts as a valve, closing the trachea while food is being swallowed.

E. Larynx: voice box (Adam's apple), upper 2 inches of trachea, just below the epiglottis. This is part of the airway system, and includes the vocal cords.

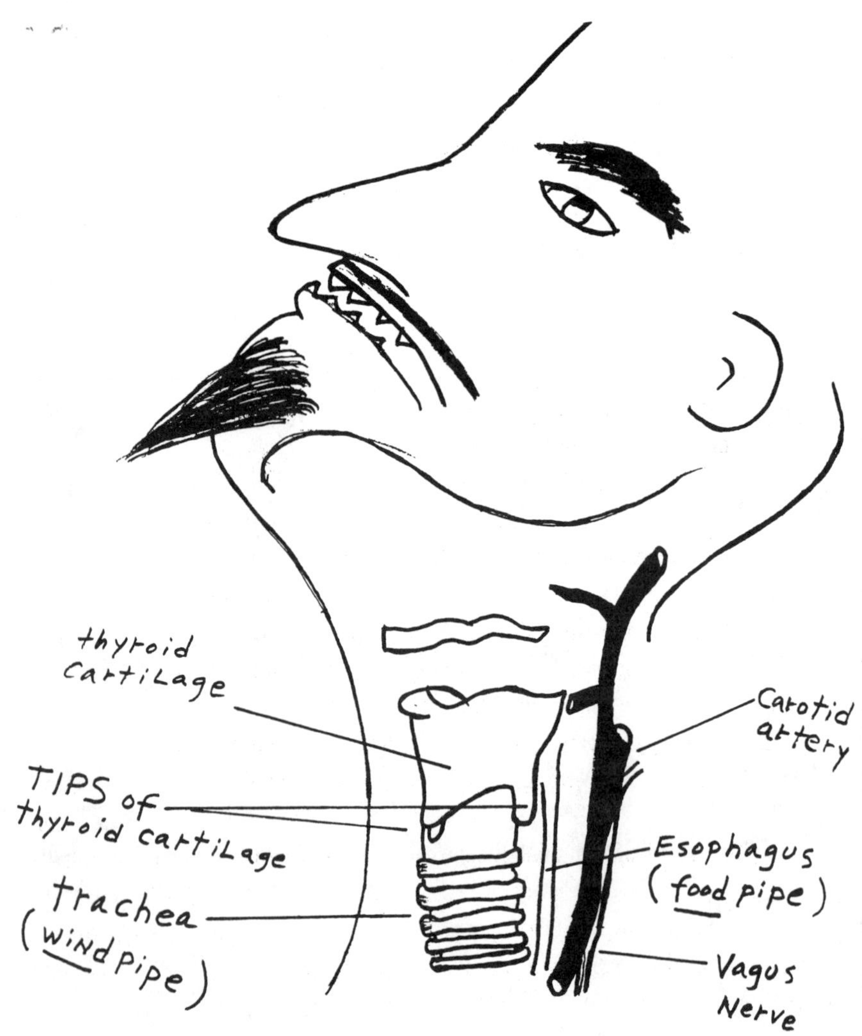

[Note. If food gets caught in the esophagus, the victim can breath. However, if food gets caught in the trachea, you must get it out pronto, because the victim can't breath. This is done by giving abdominal thrusts below the bottom edge of the rib cage, slightly above the navel. Standing behind, put the side of your fist in this spot, grasp this fist with the other hand, and press it into the victim's abdomen with quick, upward thrusts until you dislodge the piece of food. This is called the Heimlich Maneuver, and it squeezes air up and shoots out the food piece. But please buy a whole book on first aid, or take a Red Cross course, because this only covers how to give the HM to conscious adults, and it only covers one, single possibility (out of many) that can cause death in every day circumstances.]

4. Continuing. Observing the diagram, you want to direct your deadly blow against the trachea and the thyroid cartilage. You want to collapse the trachea, to flatten it. If you can do this, the mucous in the throat will act as a seal, and cut off the air. Note also the carotid artery and the vagus nerve. The carotid arteries carry oxygenated blood up to the brain, and when this is cut off the assailant goes unconscious. (This is the famous 'choke hold' some police agencies allow, but it has been deadly, especially when the officer doesn't use it right).

Anyway, it takes between five and fifteen seconds to get the guy unconscious. But in a death struggle, you want that artery either severed, or squeezed till he expires, by strangling (from the rear) or throttling (from the front). Wait 'till he stops convulsing. This may take three or four minutes. If you don't hold it long enough, just figure he will get up and kill you. Also, the vagus nerve, which runs along the side of the carotid, controls the heartbeat. That goes out with it, by definition. Another method of choking is to use one hand (or even the thumb and forefinger), and squeeze the tips of the thyroid cartilage. This, also, will do him in.[129]

Nice reading isn't it? Pity what professional conflict resolvers have to get into. Someone said that war is hell. Not true, of course. It's life. If you follow history backwards, reading about war is a good way of learning. Interestingly, one or our earliest computers, called the 'Electronic Numerical Integrator and Computer', unveiled in 1946 and weighing 30 tons, was used to 'calculate the trajectories of artillery shells and to produce aiming tables for soldiers.'[130]

Anyway, it is nasty stuff. But think about what nuclear fission/fusion, and biotoxic and chemical weapons do, or even having to lie still under shelling. Much nastier. And what if this all saved your life, or the life of someone you love dearly, from being torn up, by an animal in the form of a human being? As they say, *c'est la guerre.* These are the conventions of war, baby.

Section F: Learning A Little About Guns

Chapter 1: A Little Firearms History

When good opinions and facts are mixed together, they combine to tell us that gunpowder was discovered by the Chinese around 1,000 A.D., and carried west by the Arabs, in the 14th century (who were the principal traders of the time.) The first European guns were in use by the second quarter of the 1400's.

Gunpowder, when it became standardized (many years after its discovery), had about 75% saltpeter, 15% charcoal, and 10% sulfur. In the early firearms, the gunpowder was poured down the barrel, and the lead ball pushed with a ramrod down on top of the powder (later they used wadding to keep it all in). At the rear of the barrel was a hole, and into the hole some priming powder was seated. When this primer was ignited by fire, it would pass through to the powder behind the lead ball, and the burning of it would create gases, forcing out the projectile.

Eventually a match was made, by taking a piece of cord and soaking it in potassium nitrate, and left to dry. This glowing match was then fitted into an apparatus with an S-shaped gear (called a serpentine), a sear, a sear-spring and a trigger. When the trigger was pressed, it would cause the sear to rotate the serpentine, which in turn would move the match toward the hole, to light the primer, and then boom.

These early pistols were called match-locks, in keeping with the description just given, and they were introduced to Japan by Portuguese traders around 1540. The Japanese used these pistols for hundreds of years, but the Europeans graduated to the wheel-lock pistol, either in the late 1400's or early 1500's.

The difference between the match-lock and the wheel-lock was that the wheel replaced the serpentine device (which held the match). The wheel in question was used in conjunction with pyrites, a common mineral which produced sparks. These sparks lit the primer instead of the glowing match. Then, later in the sixteenth century, a piece of flint was substituted for the pyrites, and in place of the wheel was a flat, steel plate, bringing us the flint-locks. With these, the flint in the hammer struck the steel plate to make the spark. We're discussing smooth-bore muskets here, which were loaded from the muzzle.

These flint-locks were used until the nineteenth century, or thereabouts; until percussion firing and the priming cap entered the scene, and then cartridges. Muskets gave way to rifles in the 19th century (though rifles date from an earlier period). A rifle has a 'rifled' bore, which means the barrel is grooved, giving the bullet spin, and therefore more accuracy. Also in the 19th century, repeating rifles came into use, although at the *beginning* of the Civil War in America, smooth-bore weapons were the issue for infantry. Breech loading rifles, whether single shot or repeating, weren't around much yet.

And from there modern small arms took off rather progressively. Today we can select from the revolver and the semi-automatic pistol for handguns, shotguns with or without pump or semi-automatic features; and long guns with bolt actions, lever actions, pump actions, or semi (or fully) automatic actions (though the latter are not sold to the public, generally). And the accuracy of today's small arms is nothing less than incredible.

Military Scout Snipers can take out an enemy at 2,000 yards. The cost of the bullet: about thirteen cents[131]. Very efficient warfare this, especially when you consider that the highly trained Marine Corps Scout Sniper was getting paid about forty three cents per hour[132], and especially if you compare it to the cost of *regular* shooting in Vietnam, where it took about 600,000 rounds for each hit on an enemy soldier.[133] And if you're interested in the cost of a bullet *wound* in the hospital of today, vis-à-vis the cost of a bullet at thirteen cents or less, it's about $230,000.[134] That figure is for one bullet wound, in the *singular.*

Chapter 2: A Few Words About Ballistics

Ballistics deals with the motion and impact of projectiles. There are three fields in ballistics:

1. Interior ballistics deals with the happenings inside the gun.
2. Exterior ballistics deals with the happenings while the bullet travels.
3. Terminal ballistics deals with what happens when the bullet strikes.

The cartridge holds the bullet. Inside the cartridge, behind the bullet, is the powder. Behind the powder is the primer. When the trigger of the gun is squeezed, the firing pin is driven into the primer, which explodes, causing a tiny flame to enter the compartment where the powder stays. Then the powder burns rapidly, building up gas pressure. This forces the bullet into the barrel.

The barrel has grooves in it, called 'rifling', as was mentioned before. When the bullet moves through the barrel it has expanded, and the groves and 'lands' (the inside of the barrel not grooved) grip the bullet, causing it to spin. This gives it stability in flight, and gives the gun accuracy.

The caliber of the bullet is determined by measuring its diameter, or the diameter of the interior of the barrel. The so-called .38 is, in actuality, .357 of an inch in diameter. In other words, the .38 is really a .35. To further confuse the situation, somewhere back in recent time the .357 Magnum came out. This cartridge has the same diameter as the .38. And the .357 will fire either .38's or .357 Magnum shells, the latter being more powerful. In other words, the .38, being a .35, is actually less powerful than the .357, which, itself, is a .35. Hope this is all perfectly clear. (Maybe you should just stick with the good old .45, a heavier slug, which is significant when it comes to penetrating the body and seeking out and destroying an organ therein[135].)

Shotguns shoot either several pellets, or a single slug. Shotgun calibers are the same (though the cartridges, or shells, are not identical), but usually go by the name 'gauge'. A twelve gauge shotgun originally was calculated as follows: The bore is the size that would fit a lead ball, where twelve of these balls would weigh one pound. In actuality, the twelve gauge is about .729 inches, but can vary according to manufacturer. A sixteen gauge is a little smaller (about .662), and so on down the line.

The 'buckshot' in a shotgun shell goes according to use. If you want to take out a human being or large animal, many people prefer the 'double ought' (00). This is a shell holding nine .33 caliber pellets (if it's a 2 3/4 inch cartridge, 15 pellets if it's a 3-inch magnum). From here you go up in numbers, where a number 2 shot would be for turkeys, number 5 for pheasants or ducks, number 8 for doves (there's many sizes, depending on what's being shot at).

Chapter 3: A Little About Shooting

The thing that determines whether you hit your target, is whether you align the barrel of the weapon properly, and how steady that barrel is until the bullet goes out, also whether or not your weapon is 'sighted in' properly. This latter term simply means that the distance of your target is taken into account. Your sights will be different for a target at 100 yards, and 2,000 yards, for the reason that the bullet drops in distance and time. Also, especially in distance shooting is the steadiness vital, as opposed to very short-range combat shooting.

When the powder burns, the weapon kicks, but this doesn't effect the accuracy. The reason for this is that the bullet exits the muzzle of the barrel *before* the kick. (Duelists of old were told to aim at the knees in order to hit their opponent in the heart. Apparently they didn't understand this basic principle.[136]) For this reason, all you have to worry about is how steady the gun is, *at the time* it's fired. If you're holding a handgun, there are usually two problems. The first problem is the natural steadiness of your arm and hand. This you can't change. If you're steady you'll be a better shot. The second problem is in the way you breathe, and the way you squeeze the trigger. Don't breath while you're squeezing the trigger (take a breath, let a little out, and hold it), and your squeeze should be so steady, that you don't even *know* when the gun will go off (again, this is relevant in the measure that distance is involved, and is less of a factor in very short range shooting).

The sight alignment is also vital. Here's the way it should look.

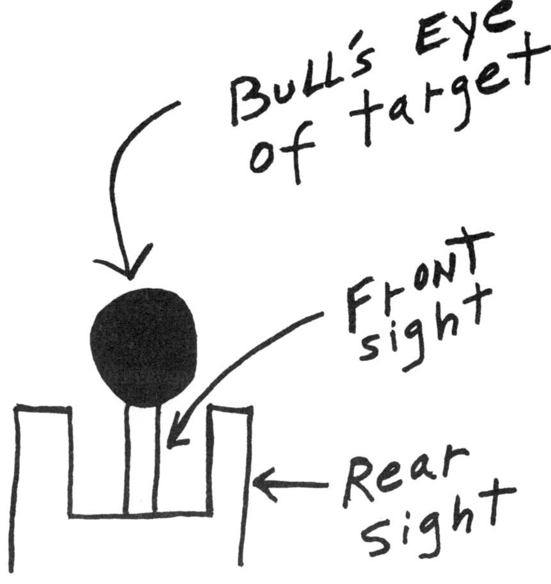

Today the trend for combat shooting is to focus on very short range shooting, because the average shoot-outs are within twenty one feet (the average *mugging* distance is three or four feet, or less). For this kind of firing, merely hold the gun out in front of you, one hand on the grip. The other hand is held palm up, which goes under the gun hand, gripping it, acting as a support. Arms outstretched (think of a revolving turret). Crouch. Move whole body as the target moves, with it acting like a revolving turret. The gun alignment in this type of shooting is much easier, since the target is very close, and therefore very big. Merely estimate where the bullet will strike (central torso of a human being). Now practice at a range.

If the reader is not into firearms (or even if he is), and wants to know more, I suggest the following video tapes, and/or books before buying a gun:

1. Deadly Weapons — firearms and firepower, $49.95
2. Deadly Effects — wounds and ballistics, what bullets do to bodies, $59.95
3. Deadly Force — firearms, self defense, and the law...Do you know when you can, and when you cannot, use a gun in self defense? $49.95. (All three of these tapes are available from Anite Company, P.O. Box 375, Pinole, CA 94564, 1-800-762-7233.)
4. Write to: P.O. Box 122, Concord, NH 03303-0122 (Lethal Force Institute, also many books available from Police Bookshelf and Armor of New Hampshire).

Chapter 4: Gun safety

"There is no such thing as a safe home, for children, with guns in that home."[137]

The above statement was made recently before the House Judiciary Subcommittee on Crime. With this kind of utter nonsense being drooled by some American citizens, I'll say a word on gun safety. For the last thirty five years I've slept with a gun, not only because I've put a lot of people behind bars, some of whom said they would come back to kill me, but because I don't want to wake up and find a mainiac in my room and be weaponless. When I go to bed I take the gun out of its place, and put it between the mattress and the box springs. When I get up in the morning, I put it back into the box.

I built a box for myself when I went into law enforcement in 1961, and the box is still in daily use. It's pine, the boards being 3/4 of an inch in actual thickness. The door of the box swings upward, hinged at the top. Through the bottom is drilled four holes, and the box is bolted onto the lower shelf of my closet, with the nuts on the inside of the box (un-notched bolt heads on outside). The latch is the kind which folds over itself, so the screws can't be removed. There are two shelves inside the box, so other valuables can be kept there also. I use a combination lock, so the kids can't get in the box.

I've never told the kids that I have a gun under my mattress at night, or what's inside that box. But I've taught them gun safety from an early age. To get them into the habit of gun safety, I'd leave an unloaded gun where they'd see it and watch them, to make sure they checked to see if it was loaded immediately upon picking it up, and didn't play with it.

I carried a gun for my entire working life. And I've kept guns in the house through 36 years of marriage, and eleven children (at one point all eleven were at home at once), and a flock of grandchildren. There's *always* kids in our house! And I've never had a gun incident.

The box probably cost me two or three dollars to build back then, including the two hinges and the latch. It measures 11 inches high, 9 1/2 inches wide, and 13 1/2 inches deep.

Please, build a box, or buy a secure gun cabinet. In today's world, you will go to *jail* if there's an accident with your gun, and it wasn't secured.

Finally, if you want to buy a gun, buy one from an established shop, from a salesman who knows guns. And take a course in shooting, and keep shooting periodically throughout the year. This is the only way you'll feel at home with your gun, and be proficient when you need it.

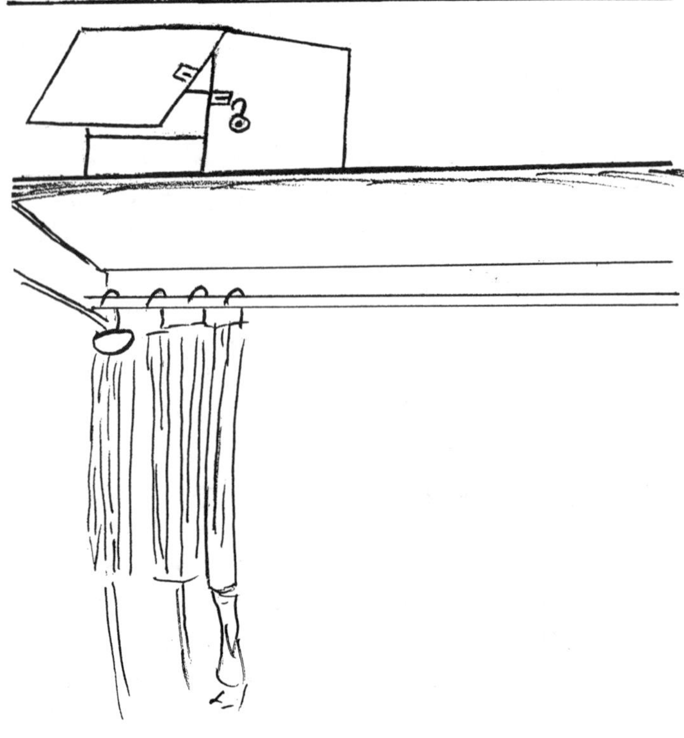

"Washington rode from the bloody fray up to the gun that a woman manned.
'Molly Pitcher, you saved the day,' he said, as he gave her a hero' s hand."

Molly Pitcher was the wife of a Revolutionary soldier, and when her husband was killed, she took his place on the battlefield, at Monmouth, June 28, 1778. The poem is called Molly Pitcher by Kate Brownlee, 1841-1914. The above is from the fifth stanza.[138]

PART FOUR: CONCLUSION TO PERSONAL AND HOME PROTECTION

Over two hundred years ago, the tyrant King George was met with resistance, *real* resistance. It erupted in April of 1775 at Lexington, and eight of the 77 Minutemen were killed. (The motivating force for the battle at Lexington was "the Crown's announced intention to disarm the colonists."[139] Please keep this in mind.)

Bunker Hill followed in June, the War was on, and our side won. It won because it was sick of the tyrant, willing to face the conflict at hand, willing to engage. And the people of this nation have enjoyed freedom ever since, by facing our enemies, dying when needed, but always by fighting.

Yet another tyrant is now upon us, a tyrant called *violence.* The Federal Center for Disease Control has called our homicide problem in America an 'epidemic'. Violence has become a *disease* in our land. Yet, out of some 25,000 murders a year, only about 300 of the perpetrators will get put on death row. And in their endless appeals, a bill of $1 million per inmate will be accrued in housing and litigation costs.[140] And to make it worse, on the federal level there hasn't been an execution since 1972, since the Supreme Court overturned the death penalty[141].

Americans aren't getting their money's worth in security. A recent study "showed that the expected time served for murder was 2.3 years; for rape, 80.5 days; for aggravated assault, 13 days; for burglary, 5 days; for auto-theft, 3.8 days".[142] A 1990 study showed similar expectations (calculated by multiplying the four probabilities — arrest, prosecution, conviction, prison — and multiplying the product by average time served for a given offense)[143].

The equation is out of balance. The system needs to be fixed. And it can be fixed by citizens getting involved. The violent offenders can be killed *before* they kill, if only the citizen will take upon himself the burden of providing for his own security. The cops and the authorities simply can't fix everything. The government isn't supposed to carry the whole burden of security.

That's not the way this country is supposed to work. When Alexis de Tocqueville came to America in the last century, to check out democracy, he worried about a deformity which is intrinsic to democracy itself. He worried that democratic man could so love his equalitarian state, he would give up his liberty to maintain it. He worried that liberty takes guts to hold onto, and that in his lust for the comforts this state can bring, the effort and vigilance to keep liberty would yield — yield to the state itself, to the experts, to the authorities — and that the people themselves would step aside, to enjoy their comforts. This is happening before our eyes. That' s the 'disease' the Center for Disease Control is talking about. Every time a person becomes oblivious to the Formula, believing he or she is protected by someone 'out there', more of his liberty is relinquished, and the liberty of all is reduced.

Yet if enough people will take part, and get involved, and learn real conflict resolution and security, the people of America can begin solving the problem. Be afraid, be alert, be prepared. The Formula really works, and it works at the local level, the only place to start in balancing the equation.

The Pollyanna syndrome in America can be broken by the people, if they will but reflect on their situation. The average grammar school child today has seen 8,000 murders on television, and been exposed to 100,000 other acts of violence on the tube, before entering the 7th grade.[144] This guzzling of blood and guts and macho worship is done by ghetto kids, all the way up the line, to kids in crystal palaces, and all too often it fulminates in killing, irrespective of milieu.[145]

The turbulent streets of our nation, already overloaded with mean people, are absorbing young minds and nervous systems and bodies weaned on frenzy. This is your lot and mine. And, unfortunately, there is no end in sight.[146]

Take heed to the dangers around you. Practice the Formula when in danger. The more you disregard it, the more defenseless you are. Good shooting, good sailing.

> There is a tide in the affairs of men
> Which, taken at the flood, leads on to fortune;
> Omitted, all the voyage of their life
> Is bound in shallows and in miseries.
> On such a full sea are we now afloat,
> And we must take the current when it serves,
> Or lose our ventures.
>
> —(Julius Caesar, Act 4, Shakespeare)

BOOK II

INTRODUCTION TO BUSINESS SECURITY

Preface

> "Law enforcement's primary concern is crime control, while the security professional's primary concern is loss prevention."[147]

'Crime control' and 'loss prevention' are two terms generally bandied about today in the West, and probably accepted by some in the business as representing the truth. As a matter of fact, law enforcement's *primary* concern is enforcing various public codes. A California cop, for example, regularly enforces the state Penal Code, the Health and Safety Code, the Vehicle Code, the Welfare and Institutions Code, and various other codes. The myriad of federal cops have their codes, as do armies of other cops, from lottery cops to fish and game cops. They all enforce the law by enforcing their codes. The proficiency of a given officer is directly related to how well he enforces his given codes, taking quantitative and qualitative considerations into account. He gets his paycheck every month as long as his proficiency reports are okay. That the crime in our country is not being 'controlled' is the great evidence of the age, but the cop still gets his paycheck, because he's doing the best he can. Every day he or she goes out and makes arrests for the same violations of the same codes, largely an after-the-fact enterprise, and the crime continues to escalate. Crime simply cannot be controlled by cops — it can only be controlled by a regime with the proper social order, which means proper social controls. The crime in America won't be controlled until the citizen stops believing the cops can do it all, and joins in the fracas.

In the private sector, the security manager is also charged with enforcing codes which deal with crime (internal criminal investigation is an integral part of many, if not most, corporate security programs). But, unlike the cops, the security manager also deals with *given entities* in before-the-fact activities. A given corporation, a given system of plants, buildings, offices, production systems, distribution systems, people, things. He does this by a fundamental method of security controls such as building design theory, entry controls, intrusion and other detection controls, and video assessment, i.e. closed circuit television; with computers playing a major role. And that is just the beginning. How much loss he 'prevents' depends on how much control he has over a given entity, which in turn depends on his own acumen and his budget. As far as 'preventing' loss, except in a qualified sense, he isn't (just as the cop is not 'controlling' crime, except in a qualified sense). The loss in the workplace due to drugs *alone* is estimated at $114 billion per year.[148] Every corporate security manager knows that until the rank and file do their part in preventing losses, 'loss prevention' is only euphonic.

Business security, corporate security, industrial security, or whatever title the security of capital or government ownership is given, is a big field. What follows is an attempt to address the *problems* of this discipline, and give the reader a road to the solutions.

A set of books covering a text-book-look at this field runs about $400 per year for annual subscription[149], (and textbooks *never* cover all of the solutions , though I *do* recommend this set for serious security executives). It includes such topics as 'security awareness of both management and employees, vulnerability training, internal/external theft and fraud, disaster control, physical security planning, investigations, ethics, locking, organization and management, standards, guard operation, security and criminal law, security and civil law, communications, alcohol and drug abuse, identification, human relationships, bombs and bomb threats, alarm sensors, proprietary data, kidnapping , extortion & terrorism, vandalism...(etc etc etc).' Since a comprehensive teaching of this subject would be impossible in a small book like this, what follows is an overview. It is hoped the reader will get from this an *understanding* of that realm, and the problems therein.

Earl's car lot:

Earl owns a used car lot, on a busy street in a city of 400,000. Adjacent to his lot, and other businesses, is a neighborhood of single family dwellings, with most of the families in the lower middle class bracket.

For several weeks Earl has experienced vandalism on some of his cars, mostly graffiti and stolen manufacturer-emblems. His lot and inventory are well-insured, but the time for repairs and parts ordering is causing him a big headache, and loss of business. So Earl calls up a friend in security management, and asks him if there is any kind of book available on used car lot security.

The friend has never seen one on that subject, but Earl is referred to various books on security, dealing with such subjects as alarms, locks, barriers, theft control, computer security, safes, guard companies, surveillance systems and a host of other subjects.

Earl gets lost in the maze of security problems and solutions, thinking 'What's all this got to do with me? All I want is to keep kids from spraying paint on my cars.'

He asks the local beat cop for any recommendations to solve the vandalism problems (the officer has paid special attention to Earl's lot lately), and he gives Earl a lesson about the crimes in question, being misdemeanors, which means the officer normally would have to witness the crime to make an arrest, or write a citation. He tells Earl that this kind of crime is very hard to control, because (like many crimes) the kids do it when nobody is watching.

The officer gives Earl some options, such as TV surveillance cameras, which could hinder some of the vandalism, but quickly mentions that the kids could just break the TV cameras. Of course there are hidden cameras on the market, and Earl could put up a sign saying 'hidden cameras' (without actually buying any). But, either way, the kids might take this as a challenge, and put bags over their faces.

The officer mentions that Earl could hire a guard company, and have a security guard on duty during the night hours, but quickly follows up with a negative note. Security guards aren't sworn

peace officers, in most cases, and when they catch the kids in the act, what then? Security guards have a different generic task from sworn officers, often acting in the prevention realm of crime and punishment rather than in an after (or during)-the-fact capacity. Recently in a California town a security guard was trying to handcuff a suspect with a gun in his hand, and shot him in the back, killing him.[150] In another recent case a security guard stabbed his supervisor to death, and was judged incompetent to stand trial, ending up in a state mental hospital for the criminally insane.[151]

These two incidents don't say much for the security guard industry, but it doesn't mean that security guards across the board are incompetent. There are security guards... and there are security guards. Some have very sophisticated training, organizations and duties. Some make arrests, just as police officers, having peace officer powers for designated duties. But some are trained primarily for simple preventive and monitoring duties, and some companies hire ex-cons. In California it can cost $100,000 to train a cop, exponentially more than your run-of-the-mill security guard.[152]

Continuing, the officer refers Earl to the security section of the yellow pages. But Earl cuts the officer off , and asks him what he would do if it was his lot. The officer thinks for a couple of minutes, and says he'd probably hire an off-duty cop. He says that this may be the least expensive way to stop the problem. He says Earl could try it for, say, two or three nights, and if the kids weren't operating during that time, Earl would only be out 'a couple hundred bucks or so'.

Earl was thinking seriously about the officer's advice, but luck was with him. That night, at just before midnight, two kids from the local neighborhood were caught in the act by the officer on the beat. Both were cited, both convicted, and both did some week-end duty picking up trash on the highway. The vandalism stopped (for several months).

Earl was elated. He knew there could be recurrences later, but he was happy that, for now, he could get back to business. And he was also happy because he learned a fundamental lesson in business security. He learned that:

1. the first thing to do is get advice, because
2. the problem in any given business could be unique to only that business, and
3. the problem might not be a permanent one, only needing a temporary solution.

And here's a good quote to keep in mind: "A successful security system should be composed of several elements, only one of which is technology."[153]

Chapter 1: Looking Into Another World

When people think of the security world, they usually think in terms of the basic security services, the low margin activities of guard companies, and various alarm, perimeter-control and institutional security patterns. They may have a general idea of risk management (which deals mostly with global loss prevention, or insurance matters); or of security management, which tends to focus on personal and property protection.

But, apart from these, there is a world in the security milieu mostly hidden to the general public, including cops and other public servants. It sometimes appears in books, though its outlines are vague, and usually somewhat fragmented or distorted.[154] It is a world where public and private sectors meet, and usually involves protection from terrorist and dissident threats, and is often manned by highly professional security people, with solid police or military backgrounds.

In this chapter, 'Looking Into Another World', I try to bring all the above into a single model, to highlight the problems of security in private business. I chose to use the literary method of hyperbole in this sketch. This seemed to be the most effective way to state the problem, since hyperbole does carry truth in its message.

The quest for security:

Nathan, making A job search, answered an ad reading: 'Wanted, master of loss prevention, for large chicken ranch and packaging plant'.

Morley, who interviewed Nathan, wore several hats, including personnel manager, chicken inventory manager, and feed-and-grain control coordinator. He also let it be known that he was 'not above' his thrice weekly stint at scooping manure 'because here, we're all just workers and brothers'.

During the conversation about the duties of the new position, Nathan learned a lot about the enterprise and its problems. The new 'loss prevention' manager would handle all security for the compound, both internal and external. This meant an outside perimeter of three square miles (over 1,8OO acres), and numerous inside perimeters for various vertical and bunker grain silos, hangars, barns, poultry houses, green houses and fields, several buildings related to the packaging plant, plus the houses and offices, and all equipment.

In addition, there was the security relating to the managers and employees (23 managers and 327 hourly, plus about 45 on-call), the hourly being heavily unionized, with a record of labor unrest.

For assistance, there was one clerk, however she was loaned intermittently to several other departments, and also took her turns at manure scooping, 'just like everyone else'.

While trying to make some security calculations in his head, about threats vis-a-vis strengths and vulnerabilities, Nathan was informed that there would be another vital duty (in his spare time, so to speak) of controlling the losses of the ranch's half million chickens. These losses had been escalating dramatically due to internal and external thefts, almost a constant at the ranch. And to make matters worse, the management wasn't certain of the quantity of the losses, due to discrepancies in the book-keeping (the book-keeper also had alternate duties in the ranch's 'feed systems dynamics' program, and, of course, the thrice weekly scooping chores). In addition, some strange new virus had been attacking the chickens, which accounted for large numbers of losses. And (perhaps the most serious problem of all) an employee in the packaging plant had apparently put a dead mouse in with a chicken, resulting in a current law-suit.

When Nathan informed Morley that he knew nothing about chicken viruses, Morley reassured him that "Anyone can learn about them".

"Would I be expected to stop these losses — which were from several causes — all at once?" asked Nathan.

"Oh, heavens no," Morley responded. "We don't expect the impossible. All we expect is continuous progress, and all progress is treated with complete equality. We here at Ivar's Chicken Ranch operate on a management theory called MBA, also known as Management by Anxiety." (Morley assured Nathan that the word 'Anxiety' was just a tongue-in-cheek thing).

"Every month you merely commit to reducing the losses of chickens by a certain number. You'd be operating on the same formula we use to control all 'negative numbers' which effect our bottom line. For example, since I am the personnel manager, I commit each month to reducing the times employees take their sick time. When all of us at the ranch do this, to reduce the 'negative numbers', you can see how it improves the general operation and the bottom line."

Morley went on, "Right now we're losing about six thousand chickens a month. So hypothetically you could commit to cutting that number down to, say, four thousand a month. Then the month after that, you make another commitment. And, of course, you make commitments on all your assigned duties, problems with labor unrest, problems with stolen equipment, and so on. See how it works?" Morley asked.

"I'm beginning to," said Nathan, who had a grasp on management theories from his college days. "Let me ask," Nathan continued,

"If I reduced the losses of chickens by 500, would this accomplishment receive the same commendation as the manager who, for instance, decreased his department's hourly chicken feeding time by .03 minutes?"

"Right!" Morley exclaimed. "You catch on fast! As I mentioned previously, equality is the name of our game here. As long as it effects the bottom line."

Nathan then asked what would happen if he didn't make his commitment.

"Good question," Morley replied. "Here's the way it works. Our staff meets twice a month. During the meetings, we interrogate each other as to why a manager didn't make his commitment. Kind of keeps the heat on, and this really effects the bottom line. Then we often team up, and help others make their commitments. So, in your case, you may have several other managers helping you in your loss prevention efforts. The communal method. We're like a family, all dedicated to the bottom line."

At this point Nathan told Morley he didn't think he would make a very good master loss prevention manager, and made a polite exit.

Morley's mistake:

Unfortunately, many people think like Morley. They get carried away with what they can expect from planning. It's been going on at least since the First World War: "It is a spirit not of providing for eventualities, but rather of attempting to preordain the future..."[155]

Since the late 1960's, or early 1970's, security did a left oblique. Not in reality, but in the mind-set of wishful-thinking utopianists. It happened in both the private and the public sector. In the *private* sector, in order to accommodate a new type of corporate security manager which was evolving, the idea of security shifted to one in which the "profit-oriented executive" would replace the manager who puts "primary stress on cost-consciousness." The focus for the security manager henceforth would be on "the prevention of loss from any cause whatsoever", in a corporation.[156]

In the *public* sector, our security policy was re-defined after Henry Kissinger's 1969 argument that the United States has "no permanent enemies" (Kissinger was at that time the assistant to the President, for National Security Affairs) and American leaders should not emphasize ideology in our foreign policy. Hans Morgenthau referred to this new position as the "ideological decontamination" of American foreign policy. About a year and a half later, President Nixon announced that, in effect, economics would replace ideology. Economics would be "the key to other kinds of power".[157]

Both of these new orders (in the private and public sectors) converge in the whole world of security, to create a potential problem. If not held in check, this idea begins a life of its own and creates a misunderstanding on the *nature* of security, as it did with Morley. This can tamper with the fundamentals of protection.

In the private sector, the reliance on a person and/or team to prevent the 'loss from any cause whatsoever' can bring on a euphoria similar to Kissinger's 'no permanent enemies' *wish.* The scope of the program which prevents the 'loss from any cause whatsoever' may sound good, but the manager in charge of such a program needs almost unlimited authority to act in various departments (accounting, personnel, operations, etc.) where his functional knowledge in those departments is less than the managers in charge. Or even worse, his functional knowledge may be non-existent. The downfall of the Soviet Union happened, at least in part, because of such managerial witchcraft. There is, in fact, a close relationship generally between prevention and totalitarianism: "...the former communist regimes were exceptionally efficient at stopping crime before it happened, simply by keeping a close watch on all citizens...The totalitarian system was very good at controlling people before they broke the law." [158]

This problem — call it the trend of self-managerialism, or just wishful thinking —is today degenerating into one of 'virtual security'. Downsizing is going on everywhere. With the virtual security trend, the security needs in a corporation are filled largely by outsiders. It is a "trend towards temporary alliances, which is finding its way into the security industry".[159] Terms such as 'outsourcing', 'partnering', 'virtual corporation', 'hybridization' are used in this new attempt at securing institutions[160]. The look is somewhat bleak for the future in institutional security (but consultants and contracting companies will likely profit). There are now 140 companies which deal in 'rent an executive'.[161]

In the public sector, the emphasis on economics, and the downplaying of ideology, causes a flood of foreign products, on the one hand, and immigrants with ulterior intentions, on the other. We speak here not of all foreign products, or all immigrants, but only those which undermine our country. This undermining of our national product and national identity can engender a togetherness euphoria, which then seeks to lessen national defense. In fact, a national security policy based on economics tends, by its very nature, to proceed the pacific way, which downplays security hardware and personnel, which is what our current policy in America is doing.

Chapter 2: Understanding That Other World

Nathan had been keeping a mental list of the things he would be held accountable for, and therefore the things that would affect his progress reports, his paycheck, his family life, and the lining of his stomach. His responsibilities at the Ranch would have been:

1. The general security of Ivar's Chicken ranch, to include:

   A. access and egress control of all employees and visitors, which meant maintaining an identification system for on-grounds checks, a key control system, which meant holding all key holders accountable, which meant that each time there was a violation, this could mean a theft, or sabotage (there was a history of union unrest), which meant there had to be some means of *enforcing* violations, of managers as well as hourly;

   B. maintaining all primary, secondary, and tertiary perimeter controls, such as fences, structural barriers, glazing material, access and egress gates, doors, screened off areas, alarm systems, lighting systems;

   C. making ongoing audits of the above systems, and evaluations of their assets and defects;

   D. understanding and updating their disaster management system, and making sure the employees kept up their drills;

   E. keeping the smoke and fire control system intact, and maintaining records on same;

   F. keeping up with employee parking problems (the ground was being cleared for a new lot, because an employee recently tripped and fell, and is suing the Ranch, and the Ranch will probably lose);

   G. maintaining a secure program for the Ranch's transportation system and staging area (many of the thefts happen here).

2. The internal security of Ivar's Chicken Ranch, to include:

   A. maintaining records on drug and alcohol abuse, and working on a program for controlling same;

   B. updating the system that deals with bomb threats (one study has it that one in every 150 threats is for real[162]), and terrorist threats;

   C. teaching security awareness;

   D. maintaining accounting and financial controls;

   E. keeping records on OSHA visits, and violations, and getting violations corrected;

F. assisting personnel with background checks and testing, and with honesty lectures, and termination's;

G. keeping updated on 'security and the criminal law', and 'security and the civil law', (there was a recent rape in the female employee's locker room, where a male employee raped and badgered a female employee, and the latter is suing the Ranch for having inadequate security for employees, which case she very likely will win);

H. bringing vandalism under control (there have been several incidents lately, but it could be minor sabotage by workers and not mere vandalism);

I. investigating thefts, which, in the words of Morley, are 'a constant' at the Ranch. Here is a real time consumer. The Sheriff's Office doesn't investigate thefts here, because they're already blown out, with too little manpower. Nathan knows how time-consuming theft investigation is, so this problem is the one that is insurmountable. When he brought this up to Morley during the interview, Morley said he would try and get Nathan 'some help' from other managers during theft investigations. But Nathan knows this is useless, because they don't know the rules of evidence, and the laws of arrest, and would likely blow any real criminal investigation.

J. maintaining a viable plan for strikes and labor disturbances.

K. keeping the Ranch's computer security updated.

L. keeping the Ranch's Product Tampering controls updated (a vital control in any business which packages food or other consumer-ingest products).

M. Keeping the Ranch's Proprietary Information Control procedure updated (Ivar's formula for its tasty chickens had been gaining some notoriety on the food-products market, and industrial espionage is on the rise everywhere[163]).

3. The planned and unplanned daily events, and minutiae of a staff manager: keeping in touch with other managers for integrated assistance of his own problems, and helping them solve their problems; keeping up on all paperwork, phone calls, on his 'commitments' (over and above his regular duties), keeping logs of trends and patterns, new developments in security, outside meetings, inside meetings; trying to get the boss to give a bigger budget to security, for more people on the payroll, maybe hiring a guard company, installing closed circuit television, and so on.

These were the reasons Nathan turned down the job. He simply wouldn't be able to cover all the bases.

Chapter 3: Learning About The Cutoff

What can be learned from the above two chapters? That there is a cutoff, a point which, when we go beyond it, we're talking apples and oranges.

The security operation for Murray's Shoe Store, run by the owner, with one salesman and a part-time clerk is the apple. A shoe store isn't a high priority for burglars. A robber might try a hold-up, but more likely he would go to a place with more cash flow. Murray is checked regularly by the local fire inspector, for keeping his fire control system working. When and if there's anything missing from the inventory or the cash register, he knows where to go for answers: it was either the clerk, the salesman, or there was a burglary. And there's no chance for labor unrest, or any of the myriad of problems at Ivar's Chicken Ranch. The owner pretty much operates as Gail did, in Chapter 5, with the 'castle mentality'. Common sense stuff really. Murray keeps his own books, and an occasional bad check or irate customer is about all he's ever had to deal with.

The orange is everything else. And everything else is a lot, because Ivar's Chicken Ranch is only a medium-sized operation. The bigger the operation, the more complicated the security system. And each security system has its own set of unique problems.

The parenthesis and perimeter concepts revisited:

For example, the security problems in manufacturing are different from those in commerce. The security problems in retail stores are different from those in hospital security. Banking security is different from campus security (students sending hate mail through the Internet on university systems are causing nightmares in campus security[164]). Hotel security is different from airport security (with both national and international terrorist threats, *pity* the poor airport security people). Transportation industry security is different from information industry security (because of computerization, and hacking and privacy problems, the information industry's security can be a brain-buster. The current estimate is that about 97% of computer crimes go undetected[165], and one of the potential threats to information security in corporations is the *government!*[166])

Convention security is different from cargo security. High rise security is different from shopping mall security. Shoplifting security is different from smash-and-grab security. Electronic transaction systems (which transfers dollars in the *trillions*) security differs from oil refinery security. AIDS in the workplace security is different from executive protection security. Cyberspace security (digital outlaws can now invade home computers through the Internet) differs from pay phone fraud security (pay phone fraud is becoming quite sophisticated).

Nuclear power station security is different from gated communities security (Gated Communities are a hot item these days, with the rise of violence. Some of them hire their own police forces, some have moats and drawbridges, or mazes of gated streets that can only be entered with key cards. One of them has even installed a 'bollard' to protect its gates. A bollard is 'an underground projectile-launching device...that shoots a 3-foot-long metal cylinder from underground into the bottom of unauthorized vehicles.' There's been some complaints from the owners of some vehicles that have 'sustained serious damage' from ignoring the sign about not entering without authorization.[167])

Museum security is different from security in the electric energy industry (electrical energy can't be stored, and it affects *everything*!) Special events security is different from public housing security. Cattle ranch security is different from chicken ranch security. (Cattle rustling is a major problem in some states, and in California they've been using micro chips under the skin of horses to identify them after being stolen[168]). Race-track security is different from cemetery security (by burying the coffin nine feet down, instead of six, thieves trying to get at a diamond-studded pistol in the coffin, gave up.[169])

Parking lot and garage security is different from maritime security. Security against armed robbery in banks is different from security for armed robbery against armored vehicles (in the former there probably won't be shooting, in the latter there probably will be). Political candidate security is different from gambling casino security. Retirement village security is different from university research lab security (who have the animal rights advocates on their backs), and these are yet again different from pleasure cruise ship security. With sarin and other such poisons as a potential threat, amusement park security is different from costume security (the Ruby Slippers worn by Judy Garland in Wizard of Oz is a hot item). And so on, through all the kinds of security entities, with a long and motley list, only a few of which are here mentioned.

All of this means that today, like it or not, specialists are needed in security. Earl, with the vandalism at his car lot, got his problem solved relatively quickly and relatively cheaply (businesses and consumers are spending $65 billion a year on security[170]). In a way, Earl was becoming a specialist in used car lot security. That point is central to this section on business security. It is the main point the reader must grasp. People make security happen.

Chapter 4: Studying Some Problems, and Some Solutions

Here are three business security problems. Also listed are the solutions the owners tried to come up with. But before you read the possible solutions, try and come up with a solution yourself, because this is what business security is all about.

1. Grocery store security. A big problem is shopping cart thefts. The carts run about $175 each, and a city of, say 150,000, may have 1,000 carts on the loose at any given time. How to solve the problem? Who solves it?
2. A private vault, advertised as 'one of the world's most secure', was beaten by three men. Two of them got into the vault by dressing up nicely, as potential clients, and being shown around by the managing director. Once inside, the two bandits held the manager at gunpoint, tied him up, tied up two security guards, and got their booty. While they were opening up the safety deposit boxes (they hit 113 out of 4,000 boxes), a third man, dressed as a security guard, turned customers away at the door. In about two hours they got off with some $38 million in cash, jewelry, bonds and gold bars. How does the company prevent another occurrence?
3. The main threat against a small Madison Avenue shop is from having a robber come in and steal the customer's money and jewelry while they're shopping. The owner has all the right security devices working already, but has to protect the customers. Armed guards would be too expensive for this small a shop. What to do?

Viewing solutions:

Here's some solutions, unlikely, possible, or probable:

1. In the first case the city (of Glendale, California) was involved, because city workers kept picking up the carts. But it was costing the city money, so the City Council wanted a measure that would force the supermarkets to bear the cost. Art Zelder, who manufactures thread fasteners, worked out a system with copper wire around the borders of the supermarket parking lots. The carts would have small receivers on them, and when they passed the copper wire, the wheels would lock. Unknown if all this worked.[171]
2. Clearly the vault company should have an I.D. system for its clients (the biometrics kind would be good, which identifies by body parts, fingerprints, retinal patterns, signature dynamics, voice, skin oil, shape of hands, palms, ears. Fancy stuff this), and should not do their sales campaigning on the vault floor. One lady lost everything she owned in this caper. Hope they were well-insured.[172]

3. The owner of the small Madison Avenue shop in New York made a store policy that all male customers had to have appointments to shop in the store, since it was that gender who did the robberies. A sign on the front read, "Men By Appointment Only". City human rights officials said she was discriminating. Probably made her stop.[173]

Chapter 5: Drawing Circles

In making an analysis of the security needs for your business, think of the circle concept. Draw an imaginary circle around each single item in your business that needs protection. Draw a circle around your safe, around your inventory, around your equipment, around your building or buildings, around a desk where an employee has compromised the books, around items that employees steal, around items that outsiders steal, around an office that has sensitive files, around a parking lot where an employee could get raped, around employees that might become dissidents, around employees that you suspect of stealing, around a counter and cash register that's been held up, around an area that caused someone to slip (who sued you).

Drawing these circles isolates the problem, and gives you a chance to contemplate what security people have to look at. Now ask yourself which of the circles (and what's inside them) can be fixed by:

1. Standard equipment, or
2. Technology, or
3. People (meaning either outside expertise, or just doing some in-house thinking), or
4. Any combination of the above.

Remember the case of Earl and his used car lot. His was a people problem. Ivar's Chicken Ranch needed a combination of all. The shopping cart problem had a technology solution. The vault problem had a combination people and technology solution. The Madison Avenue shop (if the "Men By Appointment Only" solution worked) was a people solution.

Chapter 6: Drawing Conclusions

What emerges from our little study on business security, is that often there is no easy fix. But two things stand out as good principles for business security:

1. The analysis of the cause-effect, of the problem-solution, must be made by someone who has a thorough understanding of the entity which needs protection. In the circle drawing exercise, the clerk or the janitor may be most familiar with the problem itself.

   In the smaller businesses this can be the owner or manager who solves the problem. In the larger businesses, this probably means an in-house expert, or an outside expert working closely with someone intimately familiar with the operation.

2. Achieving security in a business is very similar to any kind of problem solving. It consists of:

   A. identifying the problems,

   B. prioritizing the problems,

   C. defining the problems,

   D. planning the solutions,

   E. implementing the solutions,

   F. evaluating the operation of the plan (i.e. see if the solutions are working).

   G. If the plan is not working, re-group and find out where the mistakes are, and start fresh.

A book easily could be written on each kind of security problem. And behold! The reader has a veritable library to consult to help solve security problems, from buying the right kind of alarm[174], to sentry dogs. Merely write to one or both of the below:

1. The American Society for Industrial Security (ASIS), 1655 N. Fort Myer Drive, Suite 1200, Arlington, Virginia 22209. Ask for their Cumulative Index, or for book recommendations.
2. Butterworth-Heinemann, 313 Washington Street, Newton, MA 02158, 1-800-366-2665. Ask for their Security Book Catalog.

   Also recommended for small and medium sized businesses are the following (if any of them are out of print, maybe you can initiate a re-printing).
3. Alarm Systems & Theft Prevention — "Think Like a Thief", by Thad L. Weber, Security World Publishing Co., Inc., 2639 South La Cienega Boulevard, Los Angeles, CA 90034.

4. Stealing — How America's Employees are Stealing Their Companies Blind, by Mark Lipman, Harper's Magazine Press, NY, 1973.

5. Preventing Crime in Small Business, by Douglas L. Clark, Oasis Press, 1287 Lawrence Station Road, Sunnyvale, CA 94089 (Publishing Services, Inc., a Texas Corporation), 1984.

BOOK III

AN OVERVIEW OF PROTECTION

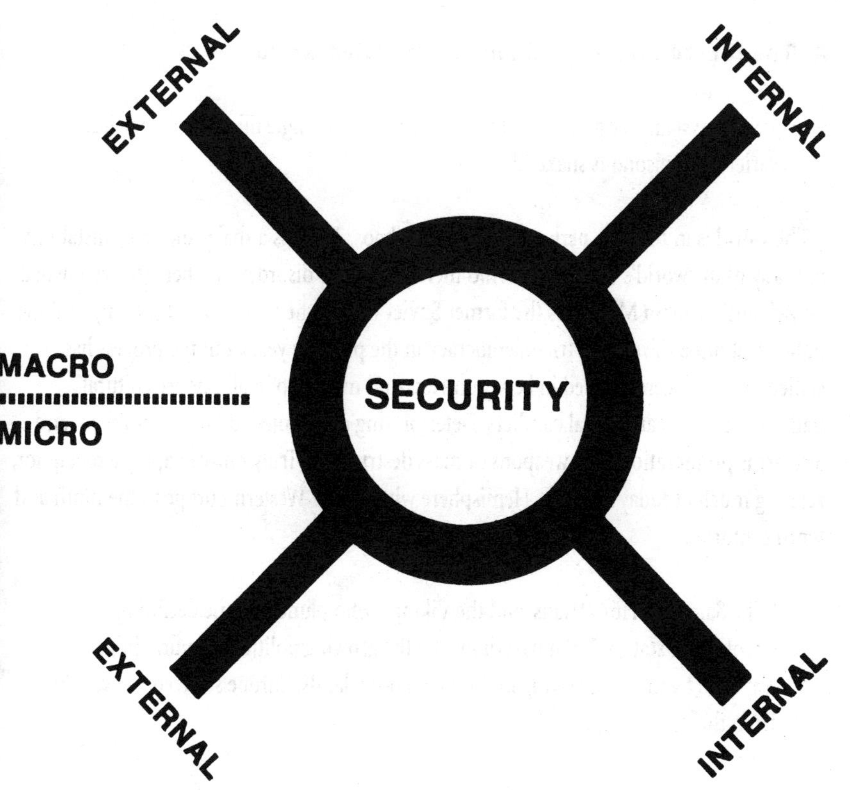
DICHOTOMY SCHEMATIC
EXTERNAL
INTERNAL
MACRO
MICRO
SECURITY
EXTERNAL
INTERNAL

Security exists in four fields of human activity:

A. At the national level our country must be prepared against a foreign (external) enemy.

B. At the national and state levels we must be prepared against internal enemies, as they could affect cities, big industry, finance, commerce.

C. At the lower levels there must be protection for the small businesses, schools and family.

D. At the lowest level there is personal protection.

Four security stories:

A. The United States: a nation in a troubled world.

> "We have slain a large dragon, but we live now in a jungle filled with a bewildering variety of poisonous snakes".[175]

The world is in turmoil, perhaps heading for chaos. There is a shaky and risky instability in many of the world's 190 nations. And there is pathetic disorder in others (the renowned French author Alain Mine calls the former Soviet Union 'the fourth world'). Many of these nations, about a third, have tried democracy in the past few years, but the process has stalled, or even been reversed in key countries. And mixed up in all this are cultural, national, religious and racial conflicts. Deteriorating economies add to the problem, and also arms proliferation with weapons of mass destruction. Truly one can apply a metaphor, relating much of today's Eastern Hemisphere with that of Western Europe in the ninth and tenth centuries.

> "The Saracens, Hungarians, and the Vikings, who plundered the declining Carolingian state, were in part drawn by the growing political vacuum, in part impelled by forces operating in their own homelands. Europe suffered grievously from their marauding..."[176]

The apex of all security is fear. Without fear, there is no foundation for security, no solid platform, nothing to build on. When there is a privation of fear at the personal level, you can lose your life. When it happens at the national level, it can be a catastrophe in terms of human and property destruction. Here are the words of a national leader who did not fear:

> "For the second time in our history, a British Prime Minister has returned from Germany bringing peace with honor. I believe it is peace for our time...Go home and get a nice quiet sleep." September 1938, 10 Downing Street, London

This fateful speech by Prime Minister Neville Chamberlain of England, was made upon his return from the infamous meeting with Hitler in Munich, Germany. There, in a ceremony which also included Prime Minister Daladier of France, these two politicians of the free world caved in to Hitler, letting him steal a section of Czechoslovakia rather than stand by their ally. What followed was a horror story for the whole world.

It began a year later in Poland. Wave upon wave of German dive bombers hit the Polish airfields, followed by fighter planes which used machine-gun fire against the hangars and barracks. In a few hours it was virtually all over, but the blitzkrieg would continue. World War II was on, but things could have been different.

In 1933, Sir Winston Churchill began making speeches to the House of Commons, warning his countrymen about the danger of Adolf Hitler. He knew that a giant airforce was being built, which would become double the size of England' s within four years. He urged that England build up her airforce, and prepare for danger. Churchill was afraid of the viper, and made estimates of this enemy, vis-a-vis the Royal Air Force. But nobody listened. The people believed the current Prime Minister Stanley Baldwin, who said there was no problem. Then, when the Polish slaughter began in 1939, the German Luftwaffe had 4,100 combat-ready planes. And no tools existed in sufficient quantity to withstand the onslaught. England had 1,900 planes, France 900, and Poland 500, but the three nations weren't ready. They weren't prepared.[177]

Fifteen million dead followed, that many more wounded. Untold misery and loss. The worst war yet. Because the Formula was not enacted. There is a lesson in this for our own nation. The world is in turmoil. And our national defense is being scaled down, rather than built up. This is happening because someone up there believes, hopes at least, that there is going to be world peace with honor. Six thousand years of war is not convincing to them.

B. A city

It was evening in the city, and the red light of the patrol car behind Morgan's car seemed to him a brilliant ruby, illuminating his whole interior and that of his expensive car. Being numbed by alcohol and drugs, his first impulse was to enjoy it, as another of the evening's pleasures. The siren's wail, however, disturbed this notion, which intensely annoyed him. And for several blocks the annoyance grew, until Morgan finally realized that the clamor was a cop car, demanding he pull over.

With a scorching anger he sharply applied his brakes, stopping abruptly, almost fully in the traffic lane. Now the top lights of the radio car were whirling red, blue and amber in a phantasmagoria of colors that made the street seem like a carnival. But neither Morgan nor the officer were seeking recreation now, as they both exited their vehicles.

"Something bothering you?" fumed Morgan with a slurring ejaculation, seeing only a badge and shiny belt buckle, now about ten feet away.

"Please step over to the other side of the police vehicle," officer Amias ordered, his voice dour.

"Why? What did you stop me for?" slurred Morgan.

"Please, over there," insisted Amias, the steady dourness already beginning to incense Morgan.

With a cocky reluctance, Morgan sauntered to the curb, his gait unsteady, tripping once on his own feet. Then he stopped and whirled around, lost his balance, then overcorrected and almost fell.

His countenance full of defiance, he now faced Amias as if they were about to begin round one in a boxing stadium, though Morgan was a heavyweight, and Amias was a light middleweight at only 153 pounds.

Amias's words seemed to be from a recording: "I stopped you for driving erratically. Please let me see your driver' s license."

After telling the officer he didn't have his license with him, Morgan was then put through a series of roadside sobriety tests. Afterwards Amias informed Morgan that he was being placed under arrest for driving under the influence, and then Morgan's constitutional rights were explained.

"Now walk to the patrol car, and place both hands on top," ordered Amias. Morgan complied.

With Morgan's hands atop the police vehicle, Amias stood behind him, and monotoned an admonition that he would now be searched.

Amias's left hand grasped Morgan's left shoulder, pulling him back slightly, and off balance. A frissoned signal of omen, almost electrical, shot into Amias's hand, something which had coursed through Morgan's body. It always happened that way. In that instant of decision, when a man's freedom was temporarily lost, and the man balked, Amias intercepted the message like a radio receiver. He pulled hard on Morgan's shoulder in a deliberated move, in order to place an arm lock on him. But in his stupor, Morgan moved surprisingly fast.

Morgan spun around and crashed a vicious fist into Amias's jaw, breaking it. But before he could take a second swing, Amias instinctively grabbed Morgan's right sleeve with his left hand, and while crushing the upper left lapel of Morgan's sport coat in his right fist, he yanked Morgan's body next to his own, burying his head under Morgan's chin. Clutching hard now, as Morgan struggled, Amias's thought process gave way to physical retaliation, as he jerked Morgan's body toward its right side, and forcefully clipped his right leg behind

Morgan's right leg, in the *osoto-gari* judo move he had used since boyhood.

Morgan's body slammed to the ground, his head thudding heavily, which baptized the pavement with a half pint of blood. The handcuffs were barely in Amias's hand when a beer bottle hit his right shoulder, and he became aware of numerous voices, screaming threats and obscenities. A brick or large rock hit the patrol car as he yanked the mike from the bracket. "Code 30! Code 30! 10-53! Garver at Hollings, north!" he tried to yell through the clenched teeth of his broken jaw.

The rear of the patrol car catapulted five feet over to the side, as a car slammed into it, the force breaking Amias's grip on the mike. Two youths exited the assault vehicle, breaking into a dead run, as the wounded officer tried to gain equilibrium. In the confusion he tried to formulate a priority upon which to act, or to just cover himself until assistance arrived. Morgan was sitting dumbly on the curb now, with one hand on the ground, his arm bracing his body. A lady with a towel was kneeling beside him, mopping his bloody head, and twenty to thirty citizens came into his view, their voices shrill with hectoring insults.

He couldn't remember if the dispatcher acknowledged his emergency call for help, and he was groping for the mike as another bottle hit his leg. He crouched beside the patrol car, to try and ascertain where the missals were coming from.

The first police assistance unit to arrive squealed to a halt near Morgan's vehicle, just as an ambulance drove up. From that point on, the incident escalated rapidly. A fire enveloped a parked car nearby, and one of the officers ran to see if there was anyone in-side. Other police cars began arriving, and just as quickly new fires broke out. The shattering of storefront glass began to add to the pandemonium. More sirens, fire units, police, ambulances.

By early morning a command post was set up in a nearby hall. Several National Guard units were responding. The first death came in mid-afternoon the next day, and a full scale riot was declared. More fires, looting, injuries, deaths.

In eleven days it was over. The inner city district in which the riot occurred was a dereliction. Every place the eye could see had burned out buildings and cars, streets filled with garbage and broken glass. Forty-one people lost their lives. Almost three hundred were injured. Over two thousand people arrested. Countless small business owners were ruined. Thousands homeless. And why?

Because of mob reaction to the injury of a drunk man who was out of control. People over-reacted to something which probably could have been resolved with a fine and a short jail sentence.

Riots are the domain not only of the government, but of the governed. Local leaders must possess fear, so they can instill this in their people. Fear of illicit drugs, fear of alcohol abuse, fear of riots and destruction. This must begin at home and in the schools. We don't live in a police state. And with communication between the leaders and the led, we won't have to.

C. A family

William and his twelve year old daughter Nancy were strolling on warm, peaceful streets. They loved their evening walk while mother did the dishes, especially during that period after the sun had dropped, and the street lights began to soften the angry night with their hovering amber. The region of the city where they lived could afford lighting with a traditional flair, and both the strollers were pleased with life.

On this night, however, something was out of place in their perfect world, as they returned from their walk. A nondescript white van was parked in a curious spot, only a hundred feet from Williams' driveway, near the Turner's house. As William and his daughter passed, he thought it might be a rental. Perhaps the neighbors were moving in some new furniture.

As the father and daughter drew adjacent to the van, however, the engine turned over, and the vehicle idled slowly at the same speed the two were walking. At first only slight discomfort chafed inside William's stomach. A glance to the driver's window showed him they were being watched, but in the darkness he couldn't see clearly inside the van.

At the next driveway, the van made a lateral turn into their path and stopped. From the side door emerged a man holding a gun, which was pointed at William. He told William to get into the van with his daughter. Nancy instinctively started to run, but William told her to do as they were told, "so they won't hurt us."

Once inside the van, the man with the gun directed the driver to drive slowly to another district. Then a third man in the back grabbed the daughter and began to molest her.

A rush of panic and nausea almost overcame William, as he moved to help his daughter, but he was struck in the head with a lead pipe, stunning him for a full five minutes. When he awakened, he was witnessing the hideous sight of bestiality on his daughter, after which one of the reprobates hacked her to death with a heavy garden tool. Then the instrument was turned on William, until he was thought dead.

Dumped into a garbage dumpster, the bodies were discovered early the next morning. William was still alive, to live the intolerable nightmare for the rest of his life. This story is taken directly from a macabre incident which happened not long ago in an American city. William had no fear, until it was too late. He was disengaged, uninhibited, smug. There was no other world for him in which the viper existed. So sure was he that their danger was not final, that he stopped his daughter from escaping.

D. A person

Even though the room was bathed bright with late morning sunshine, Florence didn't notice the grease and dirt marks on the window sill and curtains. The open window was the first thing that registered as being out of place; she had closed it earlier to thwart the morning chill. With only a malaise forming, she parted the curtains and suddenly understood. A large, jagged hole remained in the screening, which silhouetted the running body of a youth, who had dropped from the window virtually as she approached it. He was carrying something Florence could have identified a mile away: a musket which had been in the family for generations, once used by a direct ancestor in the American Revolution.

A venomous hatred surfaced at once against the thief, and her actions became spontaneous. Immediately inside the closet door was a loaded, scoped .30-06 rifle. She swung it to her shoulder, moved the safety and caught the thief perfectly in the crosshairs, just as his body was horizoned atop the fence. Rage still gripped her as her finger bore steadily down on the trigger. The squeeze was even, and the bore was still.

Almost simultaneously with the explosion came a block busting thump, which almost knocked her out of the open window. The exiting bullet tore through a stud above the pane, and the rifle crashed downward to the flowerbed. Her husband, visibly shaken, yanked her back into the room, then jumped through the window and was off in hot pursuit.

In confusion and pain Florence went into the living room and sank into an easy chair. For what seemed like two hours she sat with her eyes closed, not really thinking, slowly returning to normalcy. Then came the phone call. Her husband, Thomas, a lawyer, was calling from the Hall of Justice. A citizen just starting his car had witnessed the youth jump the fence, and knew instinctively what had happened. He idled the car behind the sprinting thief for three blocks, at which time the musket was thrown into the back of a parked pickup. The citizen was able to summon a police officer and tell what happened, after which the boy was stopped for questioning. During the field interrogation he admitted the burglary. Thomas, the husband, had lost the suspect during the chase, but when he called the police they told him to come into the station.

The musket only received a couple of nicks and scrapes. Thomas called Florence and told her to come to the Hall, and help in identifying the suspect and the old firearm. He admonished her to say nothing about the .30-06.

When back at their residence later, Thomas got out some photographs he once had used in trial. They were in color, and showed what was left of a head that got hit with a large caliber bullet. He told Florence that taking a life in a life-death struggle is one thing. But shooting a fleeing youth after a burglary is another. He asked her what she would feel like if the picture in front of her was the boy (it appeared he never had been arrested, and had a fantasy for old firearms). He had seen Thomas enter their residence with the old relic on a previous occasion. Florence, in tears, said it would haunt her for the rest of her life. She thanked her husband for the thump, and promised not to act impulsively ever again.

Fear of acting as an outlaw is vital. When this fear actuates, it will bring one to an awareness of the law, and to a healthy respect for authority. For incidents on the streets, or within the home or business, security cannot exist without this awareness and respect. Florence almost learned it too late. Had she shot the youth, she, herself, would have probably faced the judge. The crime the young man committed was a felony. But there are felonies...and there are felonies. In a modern judicial irony, the *de facto* definition of a felony sometimes turns out to be *where* and not *what.* That is, the *institution* in which the crime is punished defines the crime.

State and federal prisons house felons. County jails house misdemeanants (except for inmates awaiting transportation to prison). This certainly does not mean that all prisoners serving time in county jails committed only misdemeanors. There are inmates in county jails doing time for very serious felonies, but were, nevertheless, sentenced to jail instead of prison (the maximum for jail is usually one year).

What citizens must know is that when a criminal offense is listed in the penal code as a felony, the response of the officer or the citizen is sometimes given more liberality. The qualifier 'sometimes' here is *vital.* In the case at hand, if Florence had shot the fleeing suspect, she could have faced serious litigation, fine and incarceration for years to come.

On the other hand, had she confronted the burglar in the act, where an altercation ensued, she might have been fighting for her life. Here the incident changes rapidly and radically as to using deadly force. If great bodily harm or death could follow, then deadly force could be justified.

Fear of acting as an outlaw will bring us to awareness of the law. Being secure means protection of life and valuables. But protecting them at the price of a large legal bill, a lot of time in court, and a fine and/or incarceration isn't security, it's imprudence.

Chapter 2: Security in History, Security in America

Between six and nine thousand years ago, maybe more, maybe less, recorded history began. (The figure you choose depends on what scholar is being followed...It's interesting that some of mankind's first writings are lists of *cheese* stocks[178]).

For the sake of observing security in the context of history, to see how ours' differs from our ancestors', we shall make a simple historic draft. This draft will, out of necessity, be somewhat arbitrary. Depending on the historical thinker, the time-frame of the following blocs can be moved to or fro, and there are sub-categories and special points of significance not listed. There are, to be sure, even different opinions as to fundamentals. For example, I've listed 300 A.D. as the beginning of medievalism, which, strictly speaking, it isn't. But the seed of medievalism was there at that juncture, at about the time of Constantine's victory at the Milvian bridge (Welsh monks established parish organization in Gaul during the infancy of medievalism[179], not long after Constantine embraced Christianity). Likewise, after the 13th century, medieval Christendom was on its way out, after Philip the Fair's so-called Babylonian captivity started in 1309, and with the beginning of the humanist Renaissance.[180] Thus, for our purposes, the following sketch reasonably corresponds to that which will be found in major historical texts, some in the majority, others in the minority.

What is perhaps more problematic, is attaching a defining statement to each bloc. No historical period is one thing, and one thing only. Therefore, to attach a narrow definition to a given period is to risk another arbitration. However, like hot air, which converges with cool air to form a turbulence, the melding points are not the essential subject; rather it is the fact of the storm caused by them. If looked at in this way, we can give general definition to historical blocs to assist in understanding a matter of importance to us: the security of the past, the present, and, of course, the future.

One final problem. The history of America, as generally understood, begins with the Europeans, the English on the East Coast and the Spanish in California, and therefore — *ipso facto* — it includes the history of Europe properly so-called. And the history of Europe, by the same token, is medieval Europe, with its Judaic and Christian roots. To mention this, to use this as a basis for understanding America's security today, is not to withhold the other great histories making up America in our post-modern period. It is merely to focus directly on America's beginning vis-à-vis its current civilized form.

History in a nutshell:

1. 7,000 B.C. - 1,000 B.C. Ancient history. Urban life begins in the Middle East. The Old Law becomes known under Abraham, with the Hebrews as the Chosen People.
2. 1,000 B.C. - 300 A.D. Classical period. Lands of the Middle East unite (first under Persia, then under Hellenism), then divide with the Romans and Parthians. Christianity is born, supplanting the Old Law.
3. 300 - 1300 Medieval period. Christian civilization flourishes, drawing from the Judaic and Greco-Roman past. Church and Empire foster harmony, while feudalism and subsidiarity frustrate absolutism. Christianity wars with Islam, latter goes into decline.
4. 1300 - 1900 Modern period. Feudalism gives way to the humanist Renaissance, then the absolutism of kings. Man-centered replaces God-centered in thought and actions, leads away from the horizontal (hierarchical) to vertical (equality). Composite harmony of Christendom yields to nationalistic rivalry. Philosophy ruptures into two fields (metaphysics and science), former declines and latter flourishes. Mutation occurs in science with technology, and latter is applied to weaponry. Authority centers on rule attached to economics with latter gaining upper ground.
5. 1900 - present. Post-modern period. Equality of ideals in all of man's endeavors encourages endless opinions, factions and conflicts. Gulags, mass slaughters, gargantuan wars, and throngs of refugees follow. The idea of private ownership of property slips into notion of managerialism, with its tendencies of state central planning and collectivization, forming the base for the managerial state, which in turns begins its drift towards structuralism and deconstructionism.[181]

Security in modern history:

In the Battle of the Somme in World War I, over a million men fell dead or wounded, gutting a generation of young Europeans. Humanly speaking, the carnage was a tragedy of first proportions. Yet, in the historical dimensions of the post-modern age, it was topped by tragedies considerably worse. The 1932-1933 'terror-famine' alone, deliberately inflicted by the Communist regime in Russia upon the Ukrainians, cost ten times the deaths of the Somme battle,[182] not to mention the Nazi camps, the Soviet camps of the Gulag (an acronym for the 'islands' in the Soviet Union administered by the Main Administration of Corrective Labour Camps), and the camps of Pol Pot, which massacred up to "three million of its own people" in the 1970's.[183] These mega-tragedies, and numerous others, in the post-modern age, were the direct result of *ideas.*

As can be seen by a review of the historical draft above, it is *ideals* which move history in its travels. In medieval times, the primal concern of the populace was fear, rightly understood, which is an interior state of man. This was fear of the sin of Adam and Eve, removed by baptism; and fear of actual sin during the life, removed through the sacrament of penance. This is what medieval man believed. It is what *defines* medieval man.[184]

They were constantly defending. They were, in a way, dominated with defense, with protection. Protecting the soul against everlasting damnation by baptism, against actual sin by prayer and the sacramental life. And this defense on the interior regulated and controlled — or, it could be said, spilled over into — their outward life.

Feudalism is virtually synonymous with protection, and it has been said that fighting was synonymous with Christendom, its people fighting the devil and self on the inside, and their enemies on the outside, be they barbarians, infidels, heretics or dynastic enemies. Yet there was harmony. As a rhyme of that period went:

> "God hath shapen lives three; boor and knight and priest they be"[185]

In what modern man considers an abomination, the medieval system provided order. People of that time considered life on earth as a 'speck of light between two eternity's', with suffering and hardships an integral part (a 'vale of tears'), but with many joys also. They accepted, they worked. With some exceptions, only the nobles went to war, yet the lowest peasant could rise to the highest level, the clergy.

And when that medieval period ended, the interior of man was given free reign. Not immediately, but eventually. The sin of Adam and Eve, or Original Sin as it was called, either lost its significance or was thrown out all together. And the same with actual sin. Man was free on the inside, free to make his own inquiries and determinations as to the meaning of life. Free of fear.

The Cartesians replaced faith and fear as the starting point of life and eternity, with doubt and inquiry. And from there the theories grew to a veritable smorgasbord of choices, each philosophy seeming to divide and mutate at the same time. Many of these theories conflicted with each other, but all conflicted with the faith of yore.

Battles against the barbarians and the infidels yielded to the Wars of Religion and the Thirty Years War, becoming each time more bloody and extensive. And these in turn yielded to total war, and finally to absolute war, impossibly convoluted and nefariously vicious, using the most malignant agents imaginable for killing.

Combat, as the obligation of a privileged elite, yielded to the *levee en masse* (general conscription) in the 19th century, followed by the 20th, where all men were obliged to fight. And on the inside, each new novelty in philosophy competed with the one to either side. Internally a kaleidoscope of beliefs replaced the unity that formed Christendom, yielding to a world of competition, competition in the proposals of truth, and in manners of ruling, all negating the old.

After 1789, the act of regicide was performed, with the hope of killing not only the king, but the idea of superiority in principle, replacing it with equality in the ruling process. This murdering of the king and notion of king, was supposed to bring peace, much like the meeting at Munich many years later:

> "Everything announces an age in which that madness of nations, war, will come to an end." [186]

Thereafter, the rule would come not from feats of arms or ancestry, but from those who bought and sold, the bourgeois.[187] But in this process of change, rule by the buyers and sellers would itself give 'way again, to rule by the managers, in Russia, then Germany, and even in the United States.[188]

In 1917 this killing off of ideas was extended from the political to the economic sphere. Equality in rule was not enough; there had to be equality in the distribution of goods, through management of resources, rather than ownership:

> "The theory of Communists may be summed up in the single sentence: Abolition of private property." Section One, Communist Manifesto.

The movers of the new order hungered and thirsted for security, not out of healthy fear of an enemy, but through abolition of all traditional institutions, including the family, where enemies would then disappear altogether:

> "(Communism demands) Abolition of the family!...In proportion as the antagonism between classes within the nation vanishes, the hostility of one nation to another will come to an end." —Section Two, The Communist Manifesto.

Without the inequality of monarchical rule, an ideological enemy was slain. And without inequality of ownership and its sequel, inequality of the distribution of goods, yet another enemy fell. With political equality, with economic equality, there would be no need for fear, except fear of those who doubted the new way, those who would obstruct the new order. The medieval idea of class harmony, lingering for awhile, falls and is crushed beneath the new belief of class disharmony, which would lead to peace through synthesis.

Security in America:

Thomas Hobbs was more inclined toward absolute rule, and Jean Jacques Rousseau toward absolute freedom. While both men influenced all of modern western political thinking, including that of America, our Founding Fathers opted on the large scale for a federal republic (rule by representation and law), and on the small scale for freedom of thought, a freedom contained by our basic law. But the slippage towards the Roussean ideal of absolute liberty, on both fronts, is occurring or has already occurred:

> "She begged me not to kill her, I gave
> her a rose
> Then slit her throat, watched her
> shake till her eyes closed
> Had sex with her corpse before I left her
> And drew my name on the wall like
> Helter Skelter" [189]

The above lyrics are from a song which was popular in 1991. Since then, I have heard words from rap songs which, not too long ago, one would expect in such places as the south cell block at San Quentin, but hardly on public radio. I won't depict them here.

This verse, if it can be called that, and now worse ones, course through American airwaves, and into the minds of millions of its youth. And the state, in a pitiful display of impotence, can't find a way to protect these youngsters, who are killing people and themselves at an unprecedented rate.

We seem to be living in an age of defenselessness. The state, with all its power, is impotent when it comes to the meteoric rise in criminal and cultural deformities. It is unable to keep violence and filth off primetime TV and movies, and suicidal and murderous and other scandalous songs off the airways. The effect is drug proliferation, unrestrained sexual habits, and irreconcilable and increasing violence on the streets and in the homes.

This is a direct effect of a gross imbalance in the power of the state, vis-a-vis power in the private sector. Through a 'dictatorship of the judiciary' our weakened populous has been imbibed with the notion it can have and do what it wants, that it is their constitutional right. And, so far, the people have let the dictator have its way.

The Communist state believed that synthesis would yield total equality, only to find it yielded total misery. Our own state seems to be slipping further and further into the poststructuralist belief that it doesn't have to secure our structures of strength and stability, beginning with the infrastructure of the family. And in the measure this is not checked by a responsible citizenry, the protection from the public sector will continue to yield to the private (private security employment and spending is rising approximately twice as fast as public law enforcement[190]).

In a far-seeing movie production of 1962, a family is terrorized, yet the police can do little to help. The chief of police sticks his neck out, but is bullied by the system and its henchmen. Even while under fire, however, he gets a cop to help the victims, but the cop gets killed. In the final analysis, it is only the father of the family who can defend his own.[191]

This idea, of the public protective sphere yielding to the private, is happening today in the formerly communist Eastern European nations:

> "Attracting new recruits (for police agencies) has been difficult because of low wages and competition from previously unknown private security firms, which can offer higher salaries and have attracted many former secret police members."[192]

And this trend from the public to the private sector was also recently broached by a world famous war theorist:

> "The spread of sporadic small-scale war will cause regular armed forces themselves to change form, shrink in size, and wither away. As they do, much of the day-to-day burden of defending society against the threat of low-intensity conflict will be transferred to the booming security business; and indeed the time may come when the organizations that comprise that business will, like the condottiere of old, take over the state." [193]

To keep a reasonable balance in America, the citizens must learn to help himself. Just hiring more police is not the answer:

> "If you just strengthen police, you will find yourself in a society where democracy is in danger...In order to avoid a very militarized way of fighting crime, everyone needs to learn how to fight crime on their own.:[194]

The effort to 'strengthen the police' is laudable (especially in the wake of its *de-strengthening* beginning with the late president Lyndon Johnson's Commission on Law Enforcement and the Administration of Justice[195]). However, if this effort is to *only* strengthen the police, a police state may easily become its sequel. By getting involved, the citizen can help himself, and help his country — to remain free.

> "Ask no guarantees, ask for no security, there never was such an animal. And if there were, it would be related to the great sloth which hangs upside down in a tree all day..." Fahrenheit 451, by Ray Bradbury, Ballantine Books, NY, 1953, P. 140.

> "Security is mostly a superstition. It does not exist in nature...Life is either a daring adventure or nothing." The Open Door, Helen Keller, 1957, c.f. The Great Thoughts, G. Seldes, Ballantine Books, NY, 1985.

EPILOGUE

1 / 2 / 3

Recently in San Francisco a man saw what he thought was a mugging, and went to the supposed victim's defense. He was stabbed with a sharpened wood chisel and died. He gave his life, in fact, for a street hack who was selling bad dope.[196]

Also recently in Denver a woman who was running from an abductor, was assisted by a passing motorist. The abductor shot the passing motorist, who is now paralyzed from the waist down. The original victim was murdered by him.[197]

These were #3 crime incidents, but the innocent Samaritans didn't know the Formula. When they reacted to someone in distress, it was natural, spontaneous. They wanted to help, but didn't have the means. One gave his life for a guy selling "a bunk drug deal", the other is paralyzed for life for helping someone who got killed anyway.

Should they have helped? Of course. You can't stop people helping people in trouble. But you can teach people to go into dangerous situations prepared. That's the Formula.

In any of the above conflicts, #1, #2 or #3, your opponent is taking a gamble. He is gambling his resources against yours.

In a #1 conflict he is gambling you won't hold the line, that you will lose control. Once you lose control, he knows he has won. Once you lose control, he will either gloat over your disorder, or he will reap the harvest from your crossing the line into troubled waters, a #2 or a #3 conflict.

In a #2 conflict your opponent is a tyrant. He is acting against the law. He is gambling not only that his resources are better than yours, but further, he is pitting his resources against those of the state. If you go too far, and get caught, you will be in the legal hot soup with him.

In a #3 conflict your opponent is not only a tyrant, but has staked a claim on your well-being, or your very life. Or that of your loved ones. He is gambling he can escape your efficacious defense, and he is gambling he can avoid paying the legal penalty for his acts.

During every shift, every law enforcement officer in America straps on a gun. He or she knows that today, any of the three conflicts can be faced, maybe all of them, so he or she is prepared to deal with all.

When more of our citizens take on the mind-set of cops, our country will begin an assent out of its morass of violence. Cops live in reality. They know what they can and must do, and they're prepared for the worst.

A citizen has as much right to defense as a cop. When the needed mind-set of the citizen happens, what follows will be a demand for just laws for firearms.

When more of our citizens step out of their dream world, and know that defense is their job, that when the chips are down, superman isn't going to fly in and take over, our country will start its assent to greatness. If they don't, it will continue sinking into the pits of ever-increasing, tragic violence, where it heads faster each day. And if the leaders of this once great country continue the defense down-sizing, the United States will exist no longer in its present form.

Footnotes:

[1] Bachelor of Arts, Certified Protection Professional (American Society for Industrial Security)

[2] The modern view, of course, is that politics is power, rule and authority, well expressed in Robert A. Dahl's book "Modern Political Analysis", Prentice-Hall, NJ, 1963. This differs radically from the ancient view: that politics is the 'master art'. (c.f. "Politics", by Aristotle.)

[3] This film, produced in 1953 by Columbia studios, and directed by Laslo Benedek, is now available on VCR tape.

[4] Many people seem to associate the Hell's Angels with this incident, but they didn't even begin until 1950 (in Fontana, California). C.f. POST, 11-20-65, "Hell's Angels", by William Murray, P. 32.

[5] Facts about the Hollister incident of 1947 taken from: "Hell's Angels - A Strange and Terrible Saga", by Hunter S. Thompson, Random House, NY, 1966, PP. 65-69.

[6] Diablo Community, August , 1991, P. 26, "My Tale of Terror", by Linda Jones.

[7] It would be more accurate to say I simply formulated a methology that was and is in existence.

[8] San Francisco Examiner, PA15, 1-16-94, "People taking control of guns", by Christopher Hitchens

[9] San Francisco Chronicle, P. 1, 11-24-93, "Crime Is Now No. 1 Concern, State Poll Says", byline Vlae Kershner

[10] KTVU, Channel 2 News, San Francisco/Oakland, 1-26-94, 10:38pm

[11] Reported in "Crime Strike", insert in February 1994 issue of American Rifleman, with sources listed as "U.S. Department of Justice, Bureau of Justice Statistics; National Institute of Justice Victimization Study"

[12] A. The APB, Vol 19, No. 9, Sept. 1992, P. 1. (In this case the victim was an off-duty highway patrol officer).
B. The Contra Costa Times, July 1, 1992, p. 5A.
C. The Contra Costa Times, August 27, 1992, P. 4B.
D. San Francisco Examiner, July 25, 1993, P. B5.
E. San Francisco Chronicle, October 1, 1993.
F. Time magazine, September 20, 1993, P. 71
G. Time magazine, August 16, 1993, P. 44.
H. San Francisco Chronicle, 1-10-96, P. A6.
I. Contra Costa Times, January 3, 1994, P. 9A.

J. San Francisco Chronicle, P A 21, 2-12-94
K.Time, 12-4-95, P. 61

13 The Figgie Report on Fear of Crime, (New York: New American Library, 1983), quoted on P 47, "Vigilante - The Backlash against Crime in America", William Tucker, Stein and Day, Publishers, NY, 1985

14 In truth there is a bad kind of optomist. This is the kind that refuses to look at negatives. Our's is the oposite, the kind that confronts the negatives.

15 "The Face of Battle", by John Keegan, Penguin Books, NY, 1976, P. 70.

16 Escobedo vs. Illinois, and Miranda vs. Arizona were two Fifth Amendment (not being a witness against yourself) cases that restricted peace officers' interrogations.

17 Sam Smith v. Lyle Arnold, California Highway Patrol, State of California, First Doe, Second Doe, and Third Doe, 8-29-72. The Department of Justice came to my defense, and my official clearance from all wrongdoing was dated 9-15-72 (NB Sam Smith is not the real name of the plaintiff.)

18 "The Face of Battle", by John Keegan, Penguin Books, NY, 1976, P. 18.

19 "A Man For All Seasons", A play by Robery Bolt, Vintage Books, NY, 1960, P. 73; Columbia Pictures, same title, 1966, produced and directed by Fred Zinnemann, staring Robert Shaw, Orson Welles, Paul Scofield, John Hurt, Susannah York, winner of 6 academy awards including best picture, now on VCR.

20 San Francisco Chronicle, P. 1, 11-8-93, "Weird and Risky Rides on Muni", byline Aurelio Rojas

21 "High Risk", Dr. Ken Magid and Carole A. McKelvey, Bantam Books, NY, 1988, P. 5. (Note: The psychopath is defined by authors as haveing APD, or Antisocial Personality Disorder, the "AIDS...of the mental health world." I.E. It is incurable. This disease is the "chronic inability to behave in conformance with the social norms, to defer gratification, control impulses, tolerate frustration, profit from corrective experiences, or identify with others and form meaningful realationships with them. " P. 21.)

22 For a delightful book, which catagorizes sixteen tempermant types, read: "Please Understand Me", by David Keirsey and Marilyn Bates, Distributed by Prometheus Nemesis Book Company, Box 2082, Del Mar, CA 92014, 1984 Gnosology Books Ltd. For an exercise in evaluating tempermant types, see the movie (available on VCR) "Flight of the Phoenix" with James Stewart, 20th Century Fox, 1966.

23 Contra Costa Times, P. A12, "Toddlers at risk of dying in pools, new report says", 7-17-95.

[24] Although the Polly Klaas kidnaping became internationally famous, in my immediate area of California, about an hour's drive from Polly's home, there were two cases of very young girls being kidnapped at night from their homes, not long before the Klaas case.

[25] San Francisco Chronicle, P. A4, 12-7-93, "Polly Was Among 85,361 Missing Across U.S.", byline Rob Haeseler

[26] Quick-release bars are usually needed on both window and door security grating for fire and other emergencies.

[27] The gray winged or common Trumpeter (psophia crepitans) is a chicken sized bird, easily tamed, that makes a good "watchdog". c.f. San Francisco Examiner, Sunday Punch, P. 6, 2-18-90, "Wes Holland's Notebook".

[28] This is an electronic device that sounds like a ferocious dog. Sharper Image #SI600. $129.95.

[29] Insight, P. 7, 1-26-87, "Pentagon Drop-in".

[30] San Francisco Chronicle, P. A12, 2-22-88, "British Youth Evades Security at Royal Palace, AP.

[31] San Francisco Chronicle, P. A12, 2-28-94, "London Atwitter Over Burglar Who Got Into Charles' Drawers", Reuters.

[32] Contra Costa Times, 5B, 11-15-95; San Francisco Chronicle, A11, 11-11-95

[33] San Francisco Chronicle, P. B9, 9-20-93, Open Forum, "Why I Have a Gun", bylined by Martha Shelley of Oakland, CA.

[34] Clausewitz, "On War", II:2., P. 134, quoted from, "On Strategy", Harry G. Summers, Jr., P. 76, A Dell Book, NY, 1982.

[35] San Francisco Examiner, 2-22-93, P. A4, "Oakland pizza man shot down in cold blood", by Don Martinez.

[36] These data in the above two textual paragraphs were taken from the following:

A. San Francisco Examiner, P. A7, 9-30-92, "U.S. top lawyer asks action in 'crime crisis' ", byline Paul Avery.
B. Contra Costa Times, P. 12A, 9-1-92, byline Dennis Cauchon, USA Today, "Some officials decry U.S. 'prison' solution".
New York City had 2,245 murders in 1990, and 184 of our military were killed in the Gulf War (which was only ten more deaths than the homicides in Oakland, CA, for 1992, a city of less than 400,000. c.f. San Francisco Examiner, P. A16, 3-10-93).
C. San Francisco Examiner, P. A6, 4-29-91.
D. U.S. News and World Report, 3-11-91.

37 The Contra Costa Times, P. 14A, 9-23-92, "'Knockout' game blamed in MIT student' s slaying", byline Tony Rogers, AP

38 A. San Francisco Chronicle, 3-29-93, P. A16, "Teacher Envisions Thousands of Women Kick Boxing", byline Rick DelVecchio.
B. San Francisco Chronicle, 11-13-92, P. A 17, "Youngsters Learn to Fight Back", byline Melinda Sacks, East Bay Edition, Bay Area and California section.

39 Generally the martial arts are for incapacitating, not killing.

40 "Point Blank: Guns and Violence in America", Kleck, Alpine De Gruyter, NY, 1991, (see especially Chapter Four on this point.)

41 "High Risk", PP. 11-12, Dr. Ken Magid and Carole A. McKelvey, Bantam Books, NY, 1988.

42 "Weak Link: The Feminization of the American Military", Brian Mitchell, Regnery Gateway, Wn., D.C., 1989

43 Contra Costa Times, P. 6B, 7-4-93, "Wealthy businessmen see women bodyguards as status symbols in China", byline Nicholas D. Rristof, NY Times..

44 San Francisco Chronicle, 1-27-95, P. A4, "Why Men, Women React Differently", by Harry F. Rosenthal, AP

45 "He and She", Cris Evatt, MJF Books, NY, 1992

46 "Beyond the Crime Lab", by Jon Zonderman, PP. 3, 40, Wiley Science Editions, NY etc., 1990.

47 "Ideas Have Consequences", P. 168, Richard M. Weaver, The University of Chicago Press,Chicago and London,1948.

48 "The Pictoral Atlas of the Universe", by Kevin Krisciunas and Bill Yenne, Mallard Press, 1989 by Brompton Books Corp., Part Two, The Local Region of our Galaxy, P. 128.

49 In 1991 auto crash deaths were 43,536, while gunshot deaths were 38,317; c.f. San Francisco Chronicle, P A3, 1-28-94, "Shooting deaths may top traffic deaths by 2003", Washington, New York Times

50 San Francisco Chronicle, P. A17, 7-12-94, Open Forum, "Guns Save Lives", by Dr. Edgar A Suter, who is "national chairman of Doctors for Integrity in Research and Public Policy."

51 San Francisco Chronicle, P. C1, 10-30-93, Grab Bag, L.M. Boyd, "The Colonies Liked Bright Colors"

[52] San Francisco Chronicle, P. A30, 4-23-93, "Poll Shows America in Deep Dumbo", by Debra J. Saunders.

[53] San Francisco Chronicle, P. A17, 7-12-94, Open Forum, "Guns Save Lives", by Edgar A. Suter, National Chairman of Doctors for integrity in Research and Public Policy.

[54] "Dear Mom - A Sniper's Vietnam", by Joseph T. Ward, Ivy Books, NY, 1991, P. 61.

[55] Time, May 30, 1988, P. 56, Medicine.

[56] Mark S. Watson, Chief of Staff: Pre/war Plans and Preparations (U.S. Army in World War II), (Washington, D.C.: USGPO, 1950, P. 12), quoted from "On Strategy", Harry G. Summers, P. 83, A Dell Book, NY, 1982.

[57] "The Face of Battle", John Keegan, Penguin Books, 1978, P. 273.

[58] A. "A Crime of Self-Defense", George P. Fletcher, The Free Press, NY, 1988.
B. Insight, 8-15-88, "Self-defense Amid Moral Disorder", PP. 62, 63.
C. Author's warning! Lest the reader get carried away, New York law allows deadly force to be used "against an imminent robbery", [c.f. P. 25 of above reference 58(A)], which this was. A robbery is a grave criminal offense against the person of a human being, as opposed to a crime against his or her property. This gives the victim more liberality in defense, because serious injury or death is being faced. One of the four suspects admitted to a police officer at the scene of the shooting, that they were going to rob Goetz, (San Francisco Chronicle, 5-27-87) Anyone who thinks he can use deadly force on people who ask for money, or for any #1 or #2 incident, will be facing a serious criminal charge, and maybe civil charges to boot. It should also be mentioned that Goetz is currently facing ongoing civil litigation charges, a time-and money-costing-event that never seems to go away for him. (c.f. Contra Costa Times, P. B1, National and World section, "Goetz's odyssey goes on", AP story, byline Larry McShane, 6-17-93); also, Goetz eventually served 250 days in jail for the concealed weapon charge (c.f. San Francisco Chronicle, 12-27-94, P. A4, "Epilogue: 10 Years After Bernhard Goetz", by Vivienne Walt, Newsday.)

[59] San Francisco Chronicle, 10-28-94, P. 1, "Stockton Mulls Looser Gun Law", by Ann Bancroft.

[60] The Atlantic Monthly, July 1995, "The Crisis of Public Order", by Adam Walinsky, P. 53.

[61] San Francisco Chronicle, 4-3-93, P. A15

[62] Reported on PP. 34, 35, American Rifleman, July, 1992 issue, "Second Amendment Message in Los Angeles", by James Jay Baker.

[63] San Francisco Examiner, P. 1, 1-16-94, "Women taking up firearms", byline Katherine Seligman

[64] San Francisco Examiner, P. A6, 7-10-95.

[65] San Francisco Chronicle, P. A13, 11-19-93, "Weapons Stolen From Military Bases Going to Gangs, Cults", Cox News Service

[66] San Francisco Chronicle, P. A22, 9-10-93.

[67] "In The Line of Duty: The Service and Sacrifice of America's Finest", by Constance Clark, Potomac Publishing, 1989.

[68] "Swords and Hilt Weapons", Barnes and Noble, 1989, P. 111.

[69] San Francisco Chronicle, P. A18, 12-19-93, "AMA Rejects Call for Ban on Pistols", AP

[70] (A) "...the FBI Uniform Crime Reports and state and local law enforcement agency reports prove that military style semi-automatic rifles were used in less than l% of all homicides, making their prohibition virtually useless in the fight against crime." American Rifleman, P. 40, September, 1993.
(B) Mr. Tennenbaum is a former police lieutenant and was with the Institute of Criminal Justice and Criminology at the University of Maryland when his article appeared in the San Francisco Examiner in 1991. In the article he pointed out how the lower class in our society are at a disadvantage in dealing with bureaucratic procedures for legal firearm possession, hence they go completely without protection or obtain illegal weapons. c.f. San Francisco Examiner, 12-2-91, P. A19, "Two Views on Gun Control in the United States."

[71] Reader's Digest, October 1992, "Why Los Angeles Burned", by Ralph Kinney Bennett, PP. 74-80

[72] The Atlantic Monthly, July 1995, "The Crisis of Public Order", by Adam Walinsky, P. 52.

[73] Contra Costa Times, P. 17A, 6-7-92, "Second Amendment gives gun rights to individuals", byline Larry Pratt; Hot Topics, by Daniel Starer, Touchstone, NY, 1995, p. 41.

[74] San Francisco Chronicle, P. A20, 12-18-93, Forum, "A Right Defended", Jeff Chan, Mountain View, CA

[75] San Francisco Chronicle, P. A24, 11-4-93, "In Defense of Boondoggles", George Will, (this program was started 90 years ago because the military recognized civilians, recruited by themselves for war, were bad shots).

[76] "In One Day", P. 79, by Rom Parker, Houghton Mifflin Company, Boston, 1984.

[77] Washington Inquirer, P. 1, 8-11-89.

78 American Rifleman, February 1994, "Gun Control is Bad Medicine" by James Jay Baker, P 40.

79 Ibid.

80 San Francisco Examiner, P. A13, 7-17-95, Bonnie Erbe and Betsy Hart, "Pro & Con: Guns and neuroses".

81 ibid

82 Guns & Ammo, December 1995, "Concealed Carry Works", by G. Sitton.

83 ibid

84 ibid

85 ibid

86 ibid

87 U.S. Department of Justice, Statistics, Criminal Victimization in the U.S., 1988, National Crime Survey, Rep., NCJ 122024, 1990, Table 58; cited on P. 264, Table 3.58, Sourcebook of Criminal Justice Statistics, 1989, U.S. DOJ, NCJ 124224, 1990; Hindelang Criminal Justice Research Center, University at Albany, NY.

88 "Public/Private Liaison, get a Piece of the Privatization Pie", by Marty L. West, P. 54-60; "A Unified Approach to Crime Prevention?", by Mark H. Beaudry, P. 98; Security Management, March, 1993.

89 Contra Costa Times, P. 2B, 6-23-94, "Shefiff dusts off Old West methods", by Mark Dennis AP.

90 San Francisco Examiner, P. C10, 12-11-94, "N.Y. police aren't enough for the Upper East Side", by Janet Crawley, Chicao Tribune.

91 One of the most widely advertised OC sprays carries the brand name Pepper Defense, from New Port Richey, Florida. Their phone number is 1-800-856-7343. But you can just try your local gun shop, which probably carries a supply.

92 Security Management, March 1995, """Defensive Weapons Do's and Don'ts", by Christine Tatum and Thomas Whittle. This finding was listed by a former NYCP officer who has been conducting pepper spray experiments since 1990.

93 San Francisco Chronicle, P. C1, 4-2-94, The Grab Bag, by L.M. Boyd.

94 San Francisco Examiner, P. A7, 9-30-92, byline Paul Avery.

95 Time, 2-7-94, cover stories on crime in U.S.A., "Lock 'em Up!...And Throw Away the Key", Jill Smolowe, P. 54

[96] Contra Costa Times, P 4B, 1-25-94, " 'Death' fence around prison zapping birds", AP

[97] San Francisco Examiner, 11-22-93, P. 1, "Prisons erecting deadly fences", byline Steven A. Capps

[98] Washington Inquirer, P. 7, 9-10-93.

[99] San Francisco Chronicle, P. 1, 9-29-93.

[100] San Francisco Chronicle, P C1, 1-15-94, "The Grab Bag" by L.M. Boyd

[101] (A) San Francisco Chronicle, P. A24, 9-10-93.
(B) San Francisco Chronicle, P. C8, 9-30-93

[102] U.S. News & World Report, PP. 46, 47, "Rooting out the violent", 11-23-92.

[103] Quote is taken from an article titled "For Whom the Bell Curves", by Richard Lacayo, P. 66, Time, 10-24-94, which references a book by Charles Murray titled "Losing Ground".

[104] Insight, 6-15-92, P. 17, "Liberals Still Don't Get It: Handouts Aren't The Answer", by Wesle Pruden

[105] The judge' s name who authored the article is Samuel S. Leibowitz. It appeared in a magazine in the late 1950's or early 1960's, before copying machines went public. I typed out the article, but failed to denote the name and date of the magazine.

[106] Washington Inquirer, P. 5, 10-8-93.

[107] Ibid.

[108] San Francisco Chronicle, P A7, 1-22-94, National Report section, "Rules Toughened On Deadbeat Dads", Boston

[109] "The Man in the Grey Flannel Suit", 1956, Gregory Peck, 20th Century Fox.

[110] Washington Inquirer, 3-26-93, P. 1, "Rx for Crime", by Eric Yoder.

[111] Contra Costa Times, P. 14A, 10-29-95, "Cops say three-strikes law boosts wariness of suspects", byline by Doug Willis, AP.

[112] San Francisco Examiner, 12-3-94, P. A4, "Life term for $5 coke sale", AP.

[113] San Francisco Chronicle, 5-4-95, P. A23, "Bike Thief's '3 Strikes' Sentence", By Jamie Beckett.

[114] Contra Costa Times, 4-18-93, P. B1, "2 judges join ranks refusing drug cases", by Joseph B. Treaster NY Times; San Francisco Chronicle, 9-15-89, P. 1, "New Drug Law Is Backfiring, Judges Say", by Harriet Chiang, SF Chronicle staff writer.

115 San Francisco Examiner, P. 1, 9-11-94, "Prisons brace for flood of convicts", by Tupper Hull.

116 "Les Miserables", Victor Hugo, Grosset and Dunlap, NY.

117 "Hot Topics", by Daniel Starer, Touchstone, NY, 1995, P. 17.

118 "Ideas Have Consequences", P. 14, Richard M. Weaver, The University of Chicago Press, Chicago and London, 1948.

119 "The Hustler", 1961, 20th Century-Fox, Paul Newman, Piper Laurie, George C. Scott, Jackie Gleason.

120 "Duel", 1971, Universal City Studios, Inc., Directed by Steven Spielberg, starring Dennis Weaver.

121 There's an item on the market today called the 'door club' which (unlike the simple and almost useless door chain) will handle up to two tons of force, or so the advertisers state. It's available at Sears, or by calling 1-800-CLUB-321.

122 Washington Inquirer, P. 3, 7-27-92, "Shining Path Brings Revolution to L.A.", byline Andrew Dick.

123 Time, June 12, 1995, P. 57.

124 Time, April 3, 1995, PP. 26-41.

125 San Francisco Chronicle, 4-28-95, P. A29, "The Oklahoma Blame Game", By Llewellyn H. Rockwell Jr.

126 "Felton & Fowler's More Best, Worst, and Most Unusual", PP. 57, 58, Bruce Felton and Mark Fowler, Thomas Y. Crowell Company, NY, 1976.

127 "Forward Into Battle", by Paddy Griffith, Presidio Press, Novato, CA, 1992, P. 172.

128 "Secrets of Modern Professional Warriors", TRS, 2945 S. Mooney Blvd., Visalia, CA 93277, P. 19.

129 Part of the above is taken from, "How To Kill", by John Minnery, Paladin Press, Box 1307, Boulder Colorado, 1973, "Lesson Two: To Kill Unarmed", and "Appendix B: Your First".

130 "How Things Work - Computers", p. 12, Time-Life Books, Alexandria, VA 1990

131 "One Shot - One Kill", by Charles W. Sasser and Craig Roberts, Pocket Books, NY et al, 1990, P. 112.

132 "Dear Mom - A Sniper's Vietnam", By Joseph T. Ward, Ivy Books, NY, 1991, P. 146.

133 "Forward into Battle - Fighting Tactics from Waterloo to the Near Future", by Paddy Griffity, Presidio Press, Novato, CA, 1992, P. 178.

[134] San Francisco Examiner, P. A. 15, 12-25-94, "The High Cost of Gun Violence".

[135] This fact is shown most aptly in the film "Deadly Effects - wounds and ballistics, what bullets do to bodies", which is listed at the end of this section.

[136] San Francisco Chronicle, 4-9-94, P. C. 1, "The Grab Bag", by L.M. Boyd.

[137] San Francisco Chronicle, 4-1-95, P. A3, "Gun Owners Tell Tales of Self-Defense", AP

[138] "Familiar Quotations", John Bartlett, Little Brown and Company, Boston, 1946, P. 659.

[139] This fact was brought out in an article by the director of public affairs for the American Pro-Constitutional Association, John Erickson, who was responding to hysterical anti-gun sentiment in Lafayette, California. C.f. Contra Costa Times, P. 25A, 2-20-94, "Anti-gun sentiment misguided - Lafayette residents want to revoke constitutional right"

[140] Insight, P. 8, 2-12-90, "Sentences that Seldom Come to an End".

[141] San Francisco Chronicle, P A22, 1-28-94, "Habeas Corpus Hocus Pocus", Debra J. Saunders

[142] Washington Inquirer, P. 1, 3-26-93, "Rx for Crime", by Eric Yoder.

[143] "The Index of Leading Cultural Indicators", William J. Bennett, A Touchstone Book, Simon & Schuster, NY, 1994, PP 34, 35

[144] San Francisco Chronicle, P. D3, 3-5-92, "Teaching Emotional Literacy", Daniel Goleman, NY Times

[145] Time, P. 60, 2-1-93, "The Honor Roll Murder", by Sally B. Donnelly.

[146] The National Research Council recently released a report stating that the problems of the American youth are nothing less than "a human and national tragedy." And, which is worse, the Council was unable to offer solutions. The National Research Council is "the principal operating agency of the National Academy of Sciences and the National Academy of Engineering...Both groups are private, non-profit, congressionally chartered institutions that provide science and technology advice to the federal government." Contra Costa Times, P. Bl, Nation and World section, 6-23-93.

[147] Security Management, December 1993, "Public/Private Liaison - Finding Common Ground", p. 27, by Kevin A. Cassidy, Robert Brandes, and Anthony J. LaVeglia, CPP.

[148] Security Concepts, P. 1, February 1994, "Drugs In The Workplace, A $100 Billion Pricetag", By Gene Elig.

[149] The 'bible' of the security management world is titled "Protection of Assets Manual", "The one source management guide for comprehensive security", published by The Merritt Company, P.O. Box 955, Santa Monica, CA 90406

[150] Contra Costa Times, P. 10A, 12-1-93, "Man's slaying by guard treated as accidental"

[151] San Francisco Examiner, P. A3, 7-31-95, "Family man's killing leaves a huge void", by Kevin Foley.

[152] San Francisco Chronicle, P. 1, 11-9-93, "California Cops are Moving Out", by Michael Taylor.

[153] Security Management, May 1995, "Perimeter Protection - Putting Alarms in Their Place", by Henri Berube.

[154] In a 1989 publication titled "The 'Terrorism' Industry", by Edward Herman and Gerry O'Sullivan, Pantheon Books, NY, the shadow of the world I'm mentioning, appears, though, in my view, not altogether accurate.

[155] "The Face of Battle", John Keegan, Penguin Books, NY, 1976, P. 266.

[156] Security World (now Security, Cahners Publishing Company, Des Plaines, IL), March 1975, "The Security/Safety Merger", Arthur E. Torrington.

[157] "Troubled Journey", P. 238, Frederick F. Siegel, Hill and Wang, NY, 1984.

[158] San Francisco Chronicle, 12-25-93, P. 1

[159] Security Management, November 1993, P. 37, "Reengineering Security's Role", by Sherry L. Harowitz

[160] ibid

[161] CNN Business News, "Managing", with Lou Dobbs, early 1995 broadcast, for transcripts call 1-800-825-5746.

[162] San Francisco Chronicle, P. B1, 2-26-94, "The Grab Bag", L.M. Boyd

[163] Security Concepts (Terra Publishing, Inc., NY), Feb., 1994, "Industrial Espionage: The Race For The Competitive Advantage", byline J.L. Sherwood

[164] San Francisco Chronicle, P. C1, 8-5-95, "Profusion of Electronic Hate Mail Puts Schools in a Bind", by Serge F. Kovaleski, Wn. Post

[165] CNN Business News, 11-21-94, 0355, Stewart Barney.

[166] Security Management, December 1994, P. 96, Viewpoint, "Is There a New Threat to Corporate Security?", by Andre J. Delpeut.

[167] San Francisco Examiner, 3-20-94, A. 6, "Closing the gate to crime", by Adam Pertman, Boston Globe.

[168] San Francisco Chronicle, 6-15-93, P. A15, "High-Tech Way to Thwart Horse Thieves", by Dan Turner.

[169] Contra Costa Times, P. 12A, 11-14-90

[170] San Francisco Chronicle, P. A17, 1-3-95, "Business Has Key Role in Curbing Crime", by Leonard H. Roberts (president of Radio Shack and director of the National Crime Prevention Council).

[171] Contra Costa Times, P. 18A, 11-18-89.

[172] San Francisco Chronicle, 7-15-87, "Up to S38 Million Lost In London Vault Heist, " AP story.

[173] San Francisco Chronicle, P. 9, 1-3-87, "N.Y. Stores Screen Male Customers," byline Jane Gross, NY Times.

[174] Until recently, alarm users had only four options for alarm signals, which consisted of "hard-line" methods. These were, "a dial-up telephone line, a leased signal line, a multiplex line, and a derived channel". With cellular technology, however, a fifth option can be considered — the cellular transceiver. C.f. Security Management (magazine), October, 1993, P. 79, "Cellular Rings in New Alarm Options", by Michael Lebowitz.

[175] Time, 1-3-94, P 34, "To Conquer The Past", by Lance Morrow (quote is from incoming CIA Director James Woolsey, testifying before the Senate Select Committe on Intelligence. He is describing "the realities of the new world order")

[176] "Medieval Europe: A Short History", P. 119, C. Warren Hollister, John Wiley & Sons, Inc., NY, London, Sydney, 1964.

[177] "The Military History of World War II"; Vol. 6, The Air War in the West, Sept. 1939 to May 1941; by Trevor Nevitt Dupuy, Col., U.S. Army Retired; PP. 1-10; Franklin Watts, Inc., NY, 1963.

[178] "Cheese and Wine - Good Cooking/Good Eating", by Carol Truax, Ballantine Books, NY, 1975, P. 4.

[179] "The Germanic Invasions - The Making of Europe AD 400-600", by Lucien Musset, Barnes & Noble Books, NY, 1965, P. 113.

[180] c.f. "The Columbia Encyclopedia", fifth edition, Columbia University Press, 1993. After Giovanni Boccaccio wrote the Decameron in the mid fourteenth century, "the courtly themes of medieval literature began to give way to the voice and mores of early modern society." p. 318.

181 Structuralism (now often called semiotics), the theory which tries to reconstruct reality scientifically. Deconstructionism (now often called post-structuralism), the attack on all theories, including structuralism, which try to base knowledge. These are post-modern cultural revolutions which seek to liberate our instincts from any rule, control or order. "Marxist philosopher Antonio Gramsci suggested that in modern society the most promising grounds for revolutionary activity rested not in economics but in culture. The instruments were the theories of structuralism, post-structuralism, semiotics and deconstructionism emerging from academic leftists in France." Washington Inquirer, 6-14-91, P. 4, "The Decline of Higher Education", by Allan Brownfield.

182 "The Harvest Of Sorrow", Robert Conquest, NY/Oxford, Oxford University Press, 1986

183 Human Events, 2-23-85, P. 16, "'Killing Fields': Unmasked Face of Communism", by Patrick J. Buchanan

184 "Age of Faith", by Anne Fremantle and The Editors of Time-Life Books, Time Inc., NY, 1965.

185 Ibid. P. 16.

186 Jean-Paul Rabaut Saint-Etienne (French Protestant philosopher), Reflexions politiques sur les circonstances presentes; cited on P. 91, "The Experts Speak", Christopher Cerf and Victor Navasky; Pantheon Books, NY, 1984.

187 "The Mask of Command", P. 4, by John Keegan, Elisabeth Sifton Books, Viking, NY, 1987.

188 "The Managerial Revolution", James Burnham, The John Day Co., Inc., NY, 1941. Author's note: So intermixed is the public and private sector today, that tax money is given to large corporations, such as McDonalds and Gallo, to advertise in foreign countries. C.f. San Francisco Chronicle, P. A27, 7-28-95, "Slimy, Craven Toadies For Fat Cats", by Debra J. Saunders.

189 Newsweek, 4-1-91, "Violence in Our Culture", P. 49, (Geto Boys recent song "Mind of a Lunatic"), story by Peter Plagens with Mark Miller in NY, Donna Fotte and Emily Yoffe in LA., and bureau reports.

190 San Francisco Chronicle, P. A8, 1-10-94, "Crime wary U.S. Companies Are Getting Their Guard Up", AP, Chicago

191 "Cape Fear", 1962, Universal, Gregory Peck, Robert Mitchum. (Please don't confuse this with the 1991 film by the same name, starring Robert DeNiro. In the latter story, the criminal who terrorizes the family has a case for recrimination, which skews the message.)

192 San Francisco Chronicle, 12-25-93, P. 1

[193] "The Transformation of War", P. 207, Martin van Creveld, The Free Press, NY, 1991.

[194] San Francisco Chronicle, 12-25-93, P. 1

[195] c.f. "Vigilante", by William Tucker, Stein and Day, Publishers, NY, 1985

[196] San Francisco Chronicle, P. A19, 4-2-94, "A Plea for Witnesses To Samaritan's Slaying", by Kevin Leary.

[197] San Francisco Examiner, P. A10, 2-21-94, "Amateur sleuth locates body in record time", by Chris Angelo AP.